More Praise for
POWER SLEEP

"I have become obsessed with the importance of personal and organizational renewal in the face of crushing pressure that many of us experience at work. Reading Jim Maas's *Power Sleep* was a stunning experience: His highly original research and conclusions have pushed my views about renewal and vigor a giant step forward. This is a book of enormous importance."

—TOM PETERS, AUTHOR OF *IN SEARCH OF EXCELLENCE*

"Rest is the basis of dynamic activity. . . . Want to be more creative, loving, and successful? Follow Dr. Maas's powerful practical advice for doing less but accomplishing more."

—HAROLD H. BLOOMFIELD, M.D.,
AUTHOR OF *THE POWER OF 5* AND *TM*

"As the world speeds up and shrinks, physical energy and mental activity increase in importance, particularly with the drag of jet travel and 55-plus–hour workweeks. . . . Here is a handbook for successful survival."

—WILLIAM E. PHILLIPS,
FORMER CHAIRMAN AND CEO, OGILVY & MATHER

"With such fierce competition in the business world today, an executive can't afford to be less than 100 percent. In *Power Sleep*, Dr. Jim Maas reveals the simple way to get there, to get to peak performance—through sleeping right! Even if you think you're doing everything you can to get a good night's sleep, *Power Sleep* shows you you're wrong. . . . I guarantee it will forever change the way you look at sleeping."

—CHUCK LEE, CHAIRMAN AND CEO, GTE

"A fascinating and valuable account of how we spend more than a third of our lives—and how to do it better."

—AUSTIN KIPLINGER, CEO, KIPLINGER PUBLICATIONS

"Because sleep is undervalued in our society, we are truly a nation at risk. Everyone can benefit from the brilliant wisdom in Dr. Maas's *Power Sleep*. It holds the key to a healthier tomorrow."

—REBECCA SMITH-COGGINS, M.D., ASSOCIATE CHIEF OF EMERGENCY MEDICINE, STANFORD UNIVERSITY HOSPITAL

"Jim Maas's new book, *Power Sleep*, is a real shocker. It makes the convincing argument that the best way to stay awake is to get more sleep."

—SOL M. LINOWITZ, FORMER U.S. AMBASSADOR TO THE ORGANIZATION OF AMERICAN STATES AND FORMER CHAIRMAN OF THE BOARD, XEROX

"Dr. Maas, one of the nation's most dynamic speakers and educators, has provided another gold mine. *Power Sleep* is a handbook that everyone who wants to be successful should use to improve sleep and daytime performance."

—MICHAEL J. LINTNER, VICE PRESIDENT, INTERIM HR SOLUTIONS, INTERIM SERVICES, INC.

"After hearing Jim Maas speak, I made changes in my sleep habits that dramatically improved my life. Now, in *Power Sleep*, Dr. Maas drives home the point that proper rest is an essential component of a successful life. A must read."

—JIM McCANN, PRESIDENT, 1–800-FLOWERS

"*Power Sleep* is fascinating required reading for anyone who wants to become a peak performer. Dr. Maas makes it crystal clear just how important sleep is to your success."

—H.F. JOHNSON, PH.D., PRESIDENT, CONSUMER PRODUCTS (AMERICA, INDIA, ASIA-PACIFIC), S.C. JOHNSON WAX

POWER
SLEEP

The Revolutionary Program That Prepares
Your Mind for Peak Performance

DR. JAMES B. MAAS

WITH MEGAN L. WHERRY,
DAVID J. AXELROD,
BARBARA R. HOGAN, AND
JENNIFER A. BLUMIN

Quill
An Imprint of HarperCollins*Publishers*

The quote on page 99 is from *My Dear Mr. Churchill*, by Walter Graebner, published by Houghton-Mifflin, 1965.

Grateful acknowledgment is made to Penguin USA and Laurence Pollinger Limited on behalf of the Estate of Frieda Lawrence Ravagli for permission to reprint seven lines from "Shadows" from *The Complete Poems of D. H. Lawrence* by D. H. Lawrence, edited by V. De Sola Pinto and F. W. Roberts. Copyright © 1964, 1971 by Angelo Ravagli and C. M. Weekley, executors of the Estate of Frieda Lawrence Ravagli. Rights throughout the British Commonwealth are controlled by Laurence Pollinger Limited on behalf of the estate of Frieda Lawrence Ravagli. Reprinted by permission of Penguin USA and Lawrence Pollinger Limited on behalf of the Estate of Frieda Lawrence Ravagli.

A hardcover edition of this book was published in 1998 by Villard Books, a division of Random House, Inc. It is reprinted here by arrangement with Villard Books, Inc.

HarperCollins books may be purchased for educational, business, or sales promotional use. For information please write: Special Markets Department, HarperCollins Publishers Inc., 10 East 53rd Street, New York, NY 10022.

First HarperPerennial edition published 1999.

Reprinted in Quill 2001.

Library of Congress Cataloging-in-Publication Data

Maas, James B.
    Power sleep : the revolutionary program that prepares your mind for peak performance / James B. Maas. — 1st ed.
      p. cm.
    Originally published: New York : Villard, 1998.
    Includes bibliographical references and index.
    ISBN 0-06-097760-4
    1. Sleep I. Title
RA786.M23 1999
613.7'9—dc21   98-34045

10 11 12 SPS/RRD 40 39 38 37 36 35 34 33

Before making any changes in prescribed health-care regimens, make sure you consult a licensed physician. While this book provides information on sleep strategies and disorders, it is not intended to be a substitute for appropriate medical diagnosis or treatment. If you are having a persistent problem, consult your physician and/or one of the accredited sleep disorders centers listed in Appendix D.

TO NANCY, DANIEL, AND JUSTIN

CONTENTS

ACKNOWLEDGMENTS

This book is based on the diligent research of hundreds of sleep researchers who have advanced our knowledge of sleep in the last few decades more than in all of past history. Wherever possible I have tried to give credit to specific individuals for key findings. However, the society of sleep researchers is a closely knit group who share ideas and discoveries openly, working cooperatively to solve the mysteries of the night. More often than not, the sleep advice in this book is based on the shared insights of several investigators whose primary motivation is not self-advancement, but rather the betterment of the human condition. If this book is helpful, it is due to the selfless and collective wisdom of the entire sleep profession.

There are individuals to whom I am profoundly indebted. For thirty years Bill Dement has shared with me his insights on sleep research and his missionary zeal for informing the world about the central role of sleep in all human endeavors. He has guided me through the maze of the night and shed light on sleep's critical relevance to the day. He is my mentor and close friend. He introduced me to fellow sleep researchers Michel Jouvet, Thomas Roth, Charles Czeisler, Mary Carskadon, David Dinges, Curt Graeber, Mark Rosekind, Richard Coleman, Richard Ferber, Wilse Webb, Roslind Cartwright, Christian Guilleminault, Roger Broughton, J. Allan Hobson, Elliot Weitzman, Helene Porte, Milton Kramer, Michael Thorpy, John Lauber, Cheryl Spinweber, Rebecca Smith-Coggins, Greg Bassuk, and Martin Moore-Ede. Their pioneering work has been the focus of my films and presentations.

I would like to express my appreciation to Dr. Louis Munchmeyer and Dr. Mark Ivanick for their contributions to the chapters on sleeping pills and sleep disorders. I also benefited greatly from the insights and intellectual challenges of my friend and colleague the late Professor Carl Sagan, who taught me the value of healthy skepticism, and who appreciated the importance of sleep (at least for the occupants of this planet) in preparing the mind to think critically and imaginatively. He left us with dreams of extraterrestrial civilizations and a profound sense of responsibility for the preservation of our fragile planet.

Thanks are also due Caren and Roger Weiss, Enid and Jerry Alpern, Michael Lintner, Michael Lorelli, Nelson Schaenen, Jr., Faust Rossi, David Stoll, Jeremy Rabkin, Robert Jossen, Michael Greve, Michael Rosman, Mary and Bernie DePalma, David Feldshuh, Bruce Levitt, Bruce Halpern, William Lambert, David Dunning, Ken Blanchard, Tom Peters, William Phillips, Charles Lee, Fisk Johnson, David Myers, Greg Carroll, Catherine Neaher, Edward Moylan, Brian McCutcheon, Mike Schafer, Nina Lewis, Thomas Williams, Ed Larkin, Tom Larkin, Maureen Rnjak, Cathy Zambetti, Jerry Carr, Sol Linowitz, Hank Bartels,

Dwight Robinson, Heidi Robinson Seitz, Mel Brooks, Lauren Ploscowe, Janet Robinson, Ken Robinson, Harvey Sampson, Robert Cowie, Chuck Henkenius, Gabriel Richter, Phil Lempert, Robin Powers, Jerry Kaye, and Howard Howland for support and encouragement.

In all my endeavors I have had the good fortune of working closely with my assistant, Cindy Durbin, whose patience, understanding, and hard work enable me to teach, produce films, consult, fulfill speaking engagements, and write with a minimum of distraction. Scores of very talented undergraduate assistants have helped review the scientific literature and have served as production assistants, film crew members, video editors, and writers. Our programs have won prizes at more than forty film festivals and numerous other citations for outstanding contributions to education and the media.

I owe a very special debt of gratitude to my research assistants, Megan Wherry, David Axelrod, Barbara Hogan, Jennifer Blumin, Tracy Brick, Nancy DeHart, Margaret Meyer, Tara Dawood, and Brian Schilling. Their enthusiasm, inquisitive minds, hard questions, and creativity have been invaluable to me as their mentor. They have taught me more than I have taught them.

I am deeply appreciative of the thousands of students and corporate executives and employees who have responded to my films and lectures with enthusiasm and have challenged me to provide meaningful information to improve their alertness, success in life, and well-being.

My sincerest thanks to my agent, Margret McBride, who took this book under her wing, challenged me to shape it for maximum understanding and benefit, and made its publication feasible. With each iteration of chapter drafts Margret became more of a believer, following the advice on sleep strategies and changing her lifestyle. Her personal anecdotes and exclamations that "this stuff really works" were substantiated by her sustained commitment, energy, and good health. Publisher David Rosenthal of Villard accepted the manuscript with enthusiasm and provided extraordinary leadership in the publication process. If you find the book enjoyable and easy to understand, it is in large measure due to the guidance of my editor, Marysue Rucci, and my production editor, Sybil Pincus. I anticipate that they will sleep much better after having put their skilled hands on *Power Sleep*.

It is only with the love, patience, and understanding of my wife, Dr. Nancy Neaher Maas, and my sons, Daniel and Justin, that I was given the time and opportunity to write this book. Hopefully with the information contained herein the Maas family can sleep easier and longer. So join us. Good night, sleep well, sweet dreams, and have a good day—every day.

INTRODUCTION

S omeone once defined a professor as "one who talks in other people's sleep." I compound the situation: I lecture about sleep. For thirty-four years I have had the pleasure of teaching the introductory psychology course at Cornell University. Undergraduates are among the most sleep-deprived members of the population, so it's not surprising that the topic of sleep would hold a particular fascination for those awake enough to listen.

Although we spend nearly one third of our lives sleeping, most people know very little about the incredibly varied activity that occurs during the course of each night and its effect on the quality of life. I, too, was one of the "uninformed" until 1969, when I ventured to make a short film on a scientist-physician who was able to detect and "capture" dreams as they occur during sleep. Filming the work of the pioneer sleep researcher Dr. William C. Dement, director of the Sleep Laboratory at Stanford University, changed the direction of my own professional career. It took only one night, thirty years ago, to hook me for good.

It was 1:30 A.M. At midnight a college student went to bed in the sleep lab with several electrodes taped to his face and scalp. A polygraph machine continuously recorded the sleeper's eye movements and brain waves, the machine's pens sketching a physiological symphony of the night on a paper trail that would be a mile long by morning.

Before the student fell asleep his brain waves had been fast; the polygraph pens moved vigorously. Thirty minutes after sleep onset the waves were slower and eye movements had all but ceased, indicating deep sleep. But now, ninety minutes later, the pens began to move vigorously once again. The sleeper's brain was very active and his eyes were darting back and forth, as if scanning the environment. Was he awake? Definitely not. The researcher aroused his volunteer and asked, "What was going through your mind just now?" The first of the night's several dreams was duly reported.

This rhythmic pattern of sleep and dreaming repeated itself every ninety minutes throughout the night. There were periods of movement and periods of quiescence, periods of dreaming and periods of total unconsciousness, as well as dramatic changes in body temperature, respiration, heart rate, and genital activity. Observing an all-night sleep-recording session was an awakening for me. Before this night I had regarded sleep as a waste of time, little more than a dull monotonous period of unconscious inactivity, occasionally punctuated by a dream usually forgotten by breakfast time.

Seeing firsthand the complexity of the sleeper's journey through the night, and intrigued by Bill Dement's fascinating experiments, I changed my opinion of sleep and began to ponder the same questions posed by dedicated sleep researchers. Why not let the brain coast in neutral to provide a period of maximal rest? Do the different brain stages and rhythms of sleep determine how you, think, remember, plan, perform, and feel during the other two thirds of your life when you're awake? If so, how much sleep do you need to function optimally? Answers to these fundamental questions are important for everyone who lies down at least once every twenty-four hours and who wants to lead a successful life. So I began reading the sleep literature in depth.

The 1953 discovery by Eugene Aserinsky that rapid eye movements and specific changes in brain-wave activity signaled the likelihood that dreaming was occurring gave us a key to unlocking the cognitive mysteries of the night and stimulated research on all aspects of sleep. Within the last four decades sleep

> Sleep plays a major role in preparing the body and brain for an alert, productive, psychologically and physiologically healthy tomorrow.

research has gone from being practically nonexistent to occupying the full attention of more than a thousand physiologists, psychologists, and physicians, including me. I became absorbed in the mainstream of sleep research that now pours forth at a prodigious rate.

Brain scientists have been able to prove that sleep is not a passive state, but rather an elaborate activity with its own positive functions. This book is devoted to the significant body of new research that demonstrates how profoundly sleep affects the quality of our life.

For anyone who wants to be successful, sleep is a necessity, not a luxury. The conclusions presented in *Power Sleep* are based on recent studies of the neurological, chemical, and electrical activity of the sleeping brain, which show that even minimal sleep loss can have profound detrimental effects on mood, cognition, performance, productivity, communication skills, accident rates, and general health, including the gastrointestinal system, cardiovascular functioning, and our immune systems. Given the role of sleep in determining daytime functioning, most alarming to me is the current extent of sleep deprivation in our society. At least 50 percent of the American adult population is chronically sleep-deprived and a similar percentage report trouble sleeping on any given night. And it's getting worse by the decade. This devastating trend can be found throughout the industrialized world.

If we don't get adequate sleep, our quality of life, if not life itself, is jeopardized. With adequate sleep, the potential for peak performance is provided every morning. Unfortunately, in our hectic society we simply do not value sleep. This prescription for disaster must change. For my part, I have been devoting many of my waking hours for the last three decades to studying and teaching about the ineluctable relationship between adequate sleep and optimal living. My dream is that, given the proper research-based information, we can all learn to appreciate and benefit from the powerful advantage of sleep in preparing the mind and body for peak performance.

Much has been written on the architecture of the night's sleep, individual sleep requirements, sleep disorders, and how to get a good night's sleep. My own contribution has been to produce national television specials on sleep, to give presentations and seminars to corporate and lay audiences on "Power Sleep," and to be a resource for countless newspaper, magazine, radio, and television reports on various aspects of sleep and alertness. My Cornell psychology class has grown over three decades from 250 students to over 1,500 each semester, the largest single university live lecture course in the world. Devoting a considerable amount of time to lectures on sleep has undoubtedly contributed to the course's popularity.

Most of my audiences outside the classroom consist of busy business executives and professionals who are highly motivated to succeed in all aspects of life. They want to be efficient and effective and, not coincidentally, healthy and alert. They know they are often tired, but sleep has a low priority in their hectic schedules. There just isn't time. There are not enough hours in the day. That's life.

There is a better way. If you understand exactly what the brain accomplishes during various stages of a night's sleep and what your individual sleep requirement is, you're in a position to become a very different person. People who learn about sleep come to value sleep and adopt better sleep habits. After a few weeks they discover, perhaps for the first time, what it really feels like to be **fully alert all day long.** Their increased efficiency gives them enough hours in the day to work, and to play. They become better spouses, better parents, and better in their careers. They become more energetic, healthier, more successful, and happier with their lives. That's life as it should be.

Enthusiastic students, business executives, health-care providers, and other professionals who have heard my presentations have urged me to share with others the exciting new findings regarding the sleeping brain and its functions. They encouraged me to write a book that would help people understand and take full advantage of the power of sleep, with all its restorative and mind-enhancing properties. So with the assistance of my student researchers, Megan Wherry, David Axelrod, Barbara Hogan, and Jennifer Blumin, here it is.

PART I

SLEEP MATTERS

Blessings on him that first invented sleep! It covers a man thoughts and all, like a cloak; it is meat for the hungry, drink for the thirsty, heat for the cold, and cold for the hot. It is the current coin that purchases cheaply all the pleasures of the world, and the balance that sets even king and shepherd, fool and sage.

—CERVANTES, *Don Quixote*

SLEEP

CHAPTER 1

LEARNING ABOUT THE POWER OF SLEEP

HOW MUCH SLEEP DO YOU GET?

Ask this question and you'll hear some interesting answers. The prolific inventor Thomas Edison slept three or four hours at night, regarding sleep as a waste of time, "a heritage from our cave days." President Clinton grabs five to six hours. The performer Janis Joplin never wanted to sleep for fear she might miss a good party. Martha Stewart, an expert on planning good parties, only sleeps four to five hours each night. The comedian Jay Leno manages five hours and the millions of Americans who stay up to watch his late-night TV show won't get much more.

Then there are those at the other end of the sleep-length spectrum. Albert Einstein claimed he needed ten hours of sleep to function well. President Calvin Coolidge demanded eleven. Nighttime sleep wasn't adequate for Presidents Lyndon Johnson and Ronald Reagan and Prime Minister Winston Churchill. They took naps (and, incidentally, so did Edison). As Reagan half jokingly remarked to members of the press, "No matter what time it is, wake me up, even if it's in the middle of a cabinet meeting."[1]

Ask Grandma her "expert" opinion and you'll get an earful of advice on sleep needs and strategies:

> Everybody needs a good eight hours of sleep.
> A heavy meal makes you sleepy.
> Snacks before bedtime aren't good for you.
> Sleep before midnight is best.
> Early to bed, early to rise, makes a man healthy, wealthy, and
>   wise.
> Older people need less sleep.

Just a friendly warning: Grandmother psychology is sometimes on target, but not always.

Since everybody on earth sleeps at least once every twenty-four hours, we should all be experts. Knowledge about sleep, just like knowledge about nutrition and exercise, is essential to your life, for happiness, productivity, and general health. **Everyone should know exactly how much sleep he or she requires to feel wide awake, dynamic, and energetic all day long. Everyone should know the strategies and techniques for getting quality nocturnal sleep for maximum daytime performance. And everyone should know how to cope with sleep deprivation when it does occur.** But, alas, we are grossly ignorant when it comes to our own need for sleep.

In today's frenetic society people who sleep six hours or less are regarded as being tough, competitive, and ambitious. If you say you need lots of sleep you run the risk of being perceived as one who lacks what it takes to be successful. Maybe you'll even be regarded as lazy. Can people function well on six or seven hours of sleep? Or does everyone actually need eight or more hours to ensure good health and optimal daytime performance? Do men need more sleep than women? Do you need less sleep as you get older? When is the best time to exercise if you want a good night's sleep? Does a glass of wine before bedtime help you sleep better? Can you accurately assess how well you slept last night? What's the ideal bedroom temperature? Are naps good for you? Strangely enough, few of us can accurately answer even the most basic questions regarding sleep. We'll test your "sleep IQ" and your "sleep strategies" in the next chapter. Expect to fail, but that's okay. Otherwise, this book would not be necessary.

ARE YOU GETTING *ENOUGH* SLEEP?

Ask yourself:

How much sleep do I get each night during the week?
Does it differ on the weekends?
Do I fall asleep the minute my head hits the pillow?
Do I need an alarm clock to wake me up?

If you're getting less than eight hours of sleep each night, including weekends, or if you fall asleep instantly, or need an alarm clock to wake up, consider yourself one of millions of chronically sleep-deprived people—perhaps blissfully ignorant of how sleepy and ineffective you are, **or how dynamic you *could* be with adequate sleep.** We'll test your "sleep deprivation" in the next chapter. Again, expect to fail; you'll be joined by the majority of our teenage and adult population.

According to sleep experts, if you want to be fully alert, in a good mood, mentally sharp, creative, and energetic all day long, you might need to spend at least one third of your life sleeping. Over an average lifetime that's a commitment of nearly twenty-four years in bed!

Who can afford so much time asleep? Motivational speakers make big money encouraging us to spend less time sleeping and more time working. They'll try to convince you that you can condition yourself to sleep just four hours a night. Yes, you can condition yourself to wake up after four hours. But I've got news for you. There's a definite downside that you're not being told. . . . Reading this book will provide some illuminating facts that might save your career, your health, and even your life.

THE POWER OF SLEEP

Given that you might need to spend at least a third of your life sleeping, you *should* know what's going on. As I mentioned in my introduction, sleep is *not* a vast wasteland of inactivity. The sleeping brain is highly active at various times during the night, performing numerous physiological, neurological, and biochemical housekeeping tasks. These are essential for everything from maintaining life itself to reorganizing and enhancing thinking and memory. This enables us to remember the past, organize the present, and anticipate the future.

The process of sleep, if given adequate time and the proper environment, provides tremendous power. It restores, rejuvenates, and energizes the body and brain. The third of your life that you should spend sleeping has profound effects on the other two thirds of your life, in terms of alertness, energy, mood, body weight, percep-

tion, memory, thinking, reaction time, productivity, performance, communication skills, creativity, safety, and good health.

If our sleep is limited, our health and daytime potential is significantly reduced, if not destroyed. With adequate sleep and its concomitant brain activity, the world is our oyster . . . a pretty good deal for something that is enjoyable to do and doesn't take much, if any, effort!

ASLEEP IN THE FAST LANE

Before Thomas Edison's invention of the electric light in 1879, most people slept ten hours each night, a duration we've just recently discovered is ideal for optimal performance. When activity no longer was limited by the day's natural light, sleep habits changed. Over the next century we gradually reduced our total nightly sleep time by 20 percent, to eight hours per night.[2] But that's not nearly the end of the story. Recent studies indicate that Americans now average seven hours per night, approximately two and a half hours less than ideal.[3] Amazingly, and foolishly, one third of our population is sleeping less than six hours each night. Are we losing our minds?

In just the last twenty years we have added 158 hours to our annual working and commuting time—the equivalent of a full month of working hours.[4] According to Dr. William Dement, professor of medicine at Stanford University, working mothers with young children have added 241 hours to their work and commuting schedules since 1969. Those who also provide care for aging parents who may have age-related sleep problems might be doubly vulnerable to loss of sleep.

We now live in a twenty-four-hour society, a "rat race" where sleep is not valued. With heavy demands of work, household chores, parenting and family responsibilities, and a desire for social life, exercise, and recreation, four out of every ten of us are cutting back on sleep to gain time for what seems more important or interesting. This can be an extremely costly and dangerous mistake. Stop sleeping altogether and you will die. Large periods of sleep deprivation, as often occur in brainwashing of war captives or cult members, "can cause even heroically patriotic citizens to denounce their own nations and ideals, to sign patently false declarations, and to join political movements that have been lifelong anathemas to them," notes J. Allan Hobson, professor of psychiatry at Harvard Medical School.[5] People who by choice or because of work, illness, or force of circumstance go without sleep for five to ten days become irrational, paranoid, confused, and even hallucinatory.

Few of us are subjected to such extreme sleep loss. But most of us, consciously or unconsciously, occasionally if not chronically, deprive ourselves or others of adequate sleep. Can we adapt to minimal sleep without feeling drowsy and experiencing a decline in mood and performance?

On a day the White House planned to bask in good economic news, President Clinton instead exploded in anger at reporters' questions. . . . Within an hour of his comments, Clinton summoned the reporter . . . Bill Plante of CBS News, to apologize for losing his temper. Clinton said he hadn't been sleeping much since the July 17 crash of TWA Flight 800.[6]

Let's look at some statistics:

- High school and college students are among the most sleep-deprived people in our population. Thirty percent fall asleep in class at least once a week.

On November 25, 1991, when President George Bush spoke at an Ohio high school, "At least a third of the high school students were clearly asleep in the overheated auditorium. . . ."[7] If these students can't stay awake for the President, it's no wonder teachers can't keep them awake.

- Thirty-one percent of all drivers have fallen asleep at the wheel at least once in their lifetime.[8] The National Sleep Foundation reports that each year on our highways at least 100,000 accidents and 1,500 fatalities (the equivalent of four fully loaded Boeing 747 airplanes) are due to falling asleep at the wheel.[9] This is a very conservative estimate, as most states do not keep adequate statistics. The actual annual figures might be as high as 200,000 accidents and 5,000 fatalities (the equivalent of twelve fully loaded 747s). In addition to the tragic loss of lives, these accidents cost American society more than $30 billion annually.

In 1990 a high school student in New Hampshire who had been named America's Safest Teen Driver fell asleep at the wheel around 5 p.m., drifting over the yellow line into oncoming traffic. He killed himself and the nineteen-year-old female driver of another car. According to his father, "Safe driving was an obsession with him. The question of why he didn't recognize the fatigue and respond to it is something we will never know."[10]

- The transportation industry is being hit hard by the ravages of sleep deprivation on the highways, the rails, at sea, and in the air. According to the National Transportation Safety Board, "Fatigue is the No. 1 factor that detrimentally impacts the ability of pilots."[11]

In the PBS television documentary "Sleep Alert," a Boeing 747 captain noted: "It is not unusual for me to fall asleep in the cockpit, wake up twenty minutes later and find the other two crew members totally asleep."[12] In another report, "A Boeing 757 captain told how his forehead hit the control column on his approach to New York's Kennedy airport as the need for sleep became overwhelming."[13]

• Even airline passengers are not exempt from the effects of sleep deprivation. Job demands are forcing business executives and government officials to operate well beyond the design specifications of the human brain and body. They undertake exhausting schedules, whisk across multiple time zones, and work long days. Often suffering from the debilitating effects of jet lag, these people's health and performance are put in jeopardy. Dr. Martin Moore-Ede, a professor of physiology at Harvard Medical School and an expert on circadian rhythms and sleep, described President Bush's grueling schedule of sixteen-hour days on the back side of the clock during a ten-day visit to Japan:

> It is 5:30 a.m. in Washington, D.C., but [Bush] has already put in a long day in Tokyo. Suddenly, under the unforgiving eye of the TV cameras, he vomits, collapses, and slides under the table at a banquet with the Japanese Prime Minister, Kiichi Miyazawa, where Bush is the guest of honor. . . . His biological clock was still set somewhere in mid-Pacific and had not yet joined him in Japan. He became just one more victim of the human drive to reach beyond our physiological capacities.[14]

• Twenty percent of all employees work at night, and suffer disproportionately from drowsiness, gastrointestinal and cardiovascular problems, infertility, depression, and accidents. Fifty-six percent of shift workers fall asleep on the job at least once a week. The *Wall Street Journal* reported that $70 billion is lost per year in productivity, accidents, and health costs as a result of workers' inability to adjust to late-night work schedules.[15]

> For example, the near cataclysmic nuclear accidents at Chernobyl and Three Mile Island all occurred in the early-morning hours, during one of two periods in the twenty-four-hour day when we are most fatigued. The disasters all started because "nightshift workers missed or were confused by warning signals on their control panels."[16]

- Medical residents and interns are among the most severely sleep-deprived individuals. Many work more than 130 hours per week in shifts of twelve to sixty hours' duration, and every other night they are on call. They may be responsible for the care of forty to sixty patients. Sometimes mistakes are made. Fatal mistakes.

> An eighteen-year-old woman died "after a night of inattentive care by fatigued and inexperienced residents in one of New York's major teaching hospitals. . . . A Manhattan grand jury concluded that the patient had received 'woefully inadequate' care and had suffered repeated mistakes by first-year interns and second-year residents who had had little sleep."[17]

We are biologically ill-prepared to function on minimal sleep. Our prehistoric genetic blueprint for sleep has not evolved quickly enough to keep up with the pace of our frenetic society that runs twenty-four hours a day. As Dr. Moore-Ede asserts, "If we operated machinery the way we are now operating the human body, we would be accused of reckless endangerment."[18] According to recent Gallup surveys, 56 percent of the adult population now reports daytime drowsiness as a problem. The cost of sleep deprivation is nothing short of devastating in terms of wasted education and training, impaired performance, diminished productivity, loss of income, accidents, illness, the quality of life, and the loss of life. Are you victimizing yourself and endangering the welfare of your family and your career by not getting adequate sleep?

SLEEP DISORDERS

Even when we're exhausted and give in to our body's demands for rest, sleep can be elusive. Being stressed, harried, and hurried can make it difficult for us to fall asleep or stay asleep. Or we may be struggling with one of more than eighty disorders of the sleeping/waking state that have been identified by sleep researchers.

Insomnia

In 1995 a Gallup poll found that 49 percent of American adults were suffering from insomnia and other sleep-related disorders, a 15 percent increase since 1991. According to many medical specialists, sleep disorders collectively constitute the number one health problem in America.[19] The National Sleep Foundation attributes this to the increasingly frantic pace of life, work pressures, and an aging population.[20]

Sleep Apnea

Thirty million Americans suffer from sleep apnea, or temporary cessation of breathing, a potentially life-threatening disorder. If you have a serious case of sleep apnea and take a sleeping pill or drink too much alcohol on a given night, you might well induce the longest rest of all—you could die in your sleep. Astonishingly, 95 percent of people with sleep disorders are undiagnosed and untreated, and must struggle through the day feeling unmotivated and exhausted.[21]

For example, a businessman, asked to testify in a public hearing held by the National Commission on Sleep Disorders Research, described his intolerable life before finally being diagnosed and treated for sleep apnea: "I was experiencing constant daytime drowsiness. I would fall into a deep sleep for short periods during meetings, conversations, and public functions. At times, I could awaken and make a very inappropriate comment only to realize that I was commenting on a dream I had just experienced. My associates began to question my mental stability. . . . It was my practice in those days to carry a large pin or penknife with which I would stab myself in the leg, arm, or hand just to stay awake in meetings and while driving. I was removed from three successive jobs within a year and a half. My income was reduced by 85 percent and my savings were all used up."[22]

Costs of Sleep Disorders

The direct costs of sleep disorders and sleep deprivation for 1990 alone were estimated at $15.9 billion. Indirect costs, in terms of productivity and accidents, were said to be $150 billion.[23] Neither of these figures takes into account the incalculable costs of suffering, family dysfunction, and the loss of human life.

A young mother whose daughter died from sudden infant death syndrome, which is linked inextricably to sleep, stated: "The day after Christmas I found [her] dead in her bassinet. No words can adequately describe the shock, horror, and pain of a parent at such a moment. To hold the cold stiff body of your infant offspring is to receive one unexpected blow, your own future deleted. . . . To think that this repeats itself seven thousand times per year, one baby every hour . . ."[24]

Even though half of American adults have trouble sleeping,

physicians rarely ask their patients how they sleep. Less than 1 percent of case histories taken by doctors during routine physical examinations even include a mention of sleep.[25] This is alarming because so many people are suffering needlessly. Do you ever have difficulty sleeping? Might you have a sleep disorder?

SHOULD WE PLACE MORE VALUE ON SLEEP?

As a result of changing lifestyles, increased work, family, and financial pressures, and a stressed-out or aging population with a correspondingly higher incidence of sleep disorders, more than 100 million Americans are, by definition, chronically sleep-deprived. The number of Americans who report trouble sleeping has risen 33 percent in just the last five years. Half of our adult population is studying, working, parenting, and playing while exhausted. We make costly mental errors. We are accident-prone. We get sick too often. We have become a nation at risk. What's more, all technologically advanced societies are experiencing the same disastrous phenomenon, and the problem, if untreated, will grow.

We do not understand the need for sleep and the consequences of sleep deprivation. We must learn to value sleep as much as we value the importance of proper nutrition and exercise. To become peak performers we must change our habits so we can emerge from the fog of sleepiness to which we have become habituated. We must learn to "Power Sleep."

LEARNING ABOUT SLEEP

Why has there been so much ignorance about sleep? The topic is rarely taught in educational settings and until recently has not been part of the medical school curriculum. Not until 1996 did the American Medical Association recognize sleep medicine as a specialty. It is no small wonder that most of us know little about the importance of sleep, the incredibly varied activity that occurs during the course of each night, sleep disorders, and the role of sleep in determining subsequent alertness. Even sleep researchers are just beginning to fully comprehend the mysteries of sleep and its powerful consequences for the quality of life.

In *Power Sleep* I share important discoveries from sleep laboratories throughout the world, to increase your awareness of the importance of sleep; help you determine your individual sleep requirement; show you how to establish good sleep habits; improve your alertness, mood, productivity, quality of life; and possibly increase your life span.

Don't get too uptight about a little sleep loss from time to time. But if you're often sleep-deprived, feel sluggish and drowsy during the day, and are not performing at a level close to your potential, I'll try to help. By following the advice in this book you will be able to use the power of sleep to prepare your mind for peak performance. You will become a different person—the person you, your parents, your spouse, your children, and your boss always wanted you to be. You should find the material in this book interesting and full of invaluable suggestions. If not, it will put you to sleep—which is perhaps even more helpful!

CHAPTER 2

SLEEP DIAGNOSTIC TESTS

GATHERING INTERESTING INFORMATION:
FOUR DIAGNOSTIC SELF-TESTS

Preparing your mind for peak performance through better sleep requires awareness of scientific findings as well as familiarity with your own sleeping/waking behaviors. We'll begin our journey into the night by asking you to take three very short tests.

Self-test A surveys your general knowledge of sleep. The correct answers will be provided throughout the book—no fair looking ahead! Self-test B ascertains how likely it is that you are sleep-deprived. Self-test C examines your current sleep practices. Self-test D probes for problems that could indicate you have a sleep disorder. (Incidentally, the first sentence noted that there would be three tests and now I've described four. Did you catch the discrepancy? If not, you're undoubtedly sleep-deprived. Be proud of yourself that you're reading this book and gaining knowledge about the power of sleep—you've already taken the first step toward establishing a new and dynamic lifestyle.)

Self-Test A: What's My Sleep IQ?

Please indicate true or false for the following statements:

True False

☐ ☐ 1. Newborns dream less than adults.

☐ ☐ 2. Men need more sleep than women.

☐ ☐ 3. Not everyone dreams every night.

☐ ☐ 4. As you move from early to later adulthood you need less sleep.

☐ ☐ 5. By playing audiotapes during the night, you can learn while you sleep.

☐ ☐ 6. Chocolate candies provided on your hotel pillow will help you sleep better.

☐ ☐ 7. If you have insomnia at night, you should take a long nap during the day.

☐ ☐ 8. Sleeping pills are very helpful for people who have had insomnia for months.

☐ ☐ 9. Arousing a person who is sleepwalking can be very dangerous.

☐ ☐ 10. A soft mattress is better than a hard one for obtaining good sleep.

☐ ☐ 11. You are most alert when you first wake up.

☐ ☐ 12. To promote optimal sleep the best time to exercise is early in the morning.

☐ ☐ 13. A sound sleeper rarely moves during the night.

☐ ☐ 14. A boring meeting, heavy meal, or low dose of alcohol can make you sleepy, even if you're not sleep-deprived.

☐ ☐ 15. Sleep before midnight is better than sleep that begins after midnight.

Self-Test B: Am I Sleep-Deprived?

Please indicate true or false for the following statements:

True False

☐   ☐   1. I need an alarm clock in order to wake up at the appropriate time.

☐   ☐   2. It's a struggle for me to get out of bed in the morning.

☐   ☐   3. Weekday mornings I hit the snooze button several times to get more sleep.

☐   ☐   4. I feel tired, irritable, and stressed-out during the week.

☐   ☐   5. I have trouble concentrating and remembering.

☐   ☐   6. I feel slow with critical thinking, problem solving, and being creative.

☐   ☐   7. I often fall asleep watching TV.

☐   ☐   8. I often fall asleep in boring meetings or lectures or in warm rooms.

☐   ☐   9. I often fall asleep after heavy meals or after a low dose of alcohol.

☐   ☐   10. I often fall asleep while relaxing after dinner.

☐   ☐   11. I often fall asleep within five minutes of getting into bed.

☐   ☐   12. I often feel drowsy while driving.

☐   ☐   13. I often sleep extra hours on weekend mornings.

☐   ☐   14. I often need a nap to get through the day.

☐   ☐   15. I have dark circles around my eyes.

Self-Test C: How Good Are My Sleep Strategies?

Please indicate true or false for the following statements:

True False

☐ ☐ 1. I go to bed at different times during the week and on weekends, depending on my schedule and social life.

☐ ☐ 2. I get up at different times during the week and on weekends, depending on my schedule and social life.

☐ ☐ 3. My bedroom is warm or often noisy.

☐ ☐ 4. I never rotate or flip my mattress.

☐ ☐ 5. I drink alcohol within two hours of bedtime.

☐ ☐ 6. I have caffeinated coffee, tea, colas, or chocolate after 6 P.M.

☐ ☐ 7. I do not exercise on a regular basis.

☐ ☐ 8. I smoke.

☐ ☐ 9. I regularly take over-the-counter or prescription medication to help me sleep.

☐ ☐ 10. When I cannot fall asleep or remain asleep I stay in bed and try harder.

☐ ☐ 11. I often read frightening or troubling books or newspaper articles right before bedtime.

☐ ☐ 12. I do work or watch the news in bed just before turning out the lights.

☐ ☐ 13. My bed partner keeps me awake by his/her snoring.

☐ ☐ 14. My bed partner tosses and turns or kicks/hits me during his/her sleep.

☐ ☐ 15. I argue with my bed partner in bed.

Self-Test D: Might I Have a Sleep Disorder?

Please indicate true or false for the following statements:

True False

☐ ☐ 1. I have trouble falling asleep.

☐ ☐ 2. I wake up a number of times during the night.

☐ ☐ 3. I wake up earlier than I would like and have trouble falling back to sleep.

☐ ☐ 4. I wake up terrified in the middle of the night, but I don't know why.

☐ ☐ 5. I fall asleep spontaneously during the day in response to high arousal, such as when I hear a funny joke.

☐ ☐ 6. I have been told I snore loudly and stop breathing temporarily during sleep.

☐ ☐ 7. I walk or talk in my sleep.

☐ ☐ 8. I move excessively in my sleep.

☐ ☐ 9. I have hurt myself or my bed partner while I was sleeping.

☐ ☐ 10. I become very confused, afraid, and/or disoriented after sundown.

☐ ☐ 11. I cannot fall asleep until very late at night or cannot wake up in the morning.

☐ ☐ 12. I cannot stay awake early in the evening and I wake up before dawn.

☐ ☐ 13. I feel mild pain or a tingling sensation in my legs just before falling asleep.

☐ ☐ 14. I physically act out my dreams during the night.

☐ ☐ 15. I am often too anxious, depressed, or worried to fall asleep.

GUIDE TO THE DIAGNOSTIC SELF-TESTS

Self-test A reveals your general knowledge of sleep. All of the statements are false. If you scored poorly, don't worry—you've got plenty of company. Consider yourself very smart to be reading this book.

Self-test B checks to see if you are sleep-deprived. If you answered true to three or more of the fifteen items, you are probably not getting enough sleep. You are not alone. To find out more about the harm that sleep deprivation can cause, pay special attention to Chapter 4, "Sleep Need and Peak Performance."

Self-test C examines your sleep strategies. If you answered true to one or more of the questions, it is likely that at least one aspect of your lifestyle is interfering with your sleep. To learn what you can do to get a better night's sleep, turn to Chapter 5, "The Golden Rules of Sleep," and Chapter 6, "Twenty Great Sleep Strategies: How to Sleep Your Way to Success—Properly!"

Self-test D probes for problems that could indicate you have a sleep disorder. If you answered true to any of the questions, carefully read Chapter 13, "Insomnia and Beyond." Some sleep disorders are more serious than others and may require medical attention. If after further reading you decide to seek professional help, consult the list of sleep disorders centers in Appendix D.

PART II

THE POWER OF SLEEP

Come, Sleep! O Sleep, the certain knot of peace,
The baiting-place of wit, the balm of woe,
The poor man's wealth, the prisoner's release
Th' indifferent judge between the high and low.

—SIR PHILIP SIDNEY, "Astrophel and Stella"

SLEEP

CHAPTER 3

THE ARCHITECTURE AND FUNCTIONS OF SLEEP

WHAT IS SLEEP?

For centuries sleep was regarded as a quiet inert state that evolved to ensure survival. Sleeping was a good way for our primitive ancestors to avoid the dangers of wandering around in the dark, to recuperate from fatigue, and to conserve energy needed for daytime foraging. The guidelines were simple: Seek safe ground, lie down, turn off the mind, rest up, and be ready to hunt or gather at daybreak.

Dreaming was not considered part of sleep; dreams were thought to be messages sent to the sleeper from the gods, or as journeys of the soul to distant lands. Thus sleep was perceived as a state of minimal body and brain activity, the very opposite of wakefulness. It seems so simple and reasonable: two completely opposite, yet reinforcing states of existence every twenty-four hours. Although this "active state vs. passive state" notion provides a commonsense and therefore widely accepted definition of wakefulness and sleep, it is inaccurate.

Any valid definition of sleep and explanation of why we do it must be in accord with emerging facts and observations. We still need to sleep even though we have artificial illumination to counter darkness. Sleep doesn't occur in response to boredom or mental or physical fatigue. Sleep isn't necessary to conserve energy, because rest alone, without sleep, can conserve energy. Sleep is not determined by eating or by the resulting stomach vapors that Aristotle and the Greek philosophers thought cooled the heart or blocked the brain's pores. Nor is sleep determined by will. And sleep does not mean the cessation of brain activity—that happens only in animal hibernation.

Sleep cannot simply be defined by the behavioral symptoms of lying down, being still, having your eyes closed, and being unresponsive to environmental stimulation. As Harvard's Dr. J. Allan Hobson notes, such symptoms can be feigned. What you can't fake is what actually defines sleep: dramatic, measurable changes in the electrical and chemical activity of the brain.[1]

Rather than being a vast wasteland of monotonous inertness, sleep is a diverse, complex, multifaceted series of stages that make important contributions to our daytime functioning. The various stages of sleep we experience each night as our senses disengage from the environment are delineated by significant changes in brain waves, muscle activity, eye movements, body temperature, respiration, heart rate, hormonal activity, and even genital arousal. The overall level of neural activity drops by only 10 percent during sleep. In fact, the "sleeping" brain is often significantly more active than the "awake" brain. We now know that various activities of the sleeping brain play a dramatic role in regulating gastrointestinal, cardiovascular, and immune functions, in energizing the body, and in cognitive processing, including the storing, reorganization, and eventual retrieval of information already in the brain, as well as in the acquisition of new information while awake. A big job, hardly the province of a passive, inert state.

UNLOCKING THE MYSTERY OF THE NIGHT

Remarkably, we knew very little about sleep until fairly recently in the history of medicine. In 1929, Hans Berger, a German psychiatrist, used small electrodes attached to the scalp to record the continuous electrical activity of the human brain. The brain-wave recordings, called "electroencephalograms," or EEGs, indicated distinctive changes in the brain's neural activity between sleep and wakefulness. The brain waves of the waking state were fast

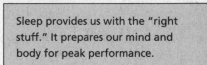

Sleep provides us with the "right stuff." It prepares our mind and body for peak performance.

in frequency (cycles per second) and low in amplitude (micro-volts of electrical discharge of brain neurons). By contrast, Berger noted that the sleeping brain was characterized by low-frequency, high-amplitude brain waves, indicating a substantial decline in neural activity. But he was unaware that sleep was not a uniform state.

In 1935 researchers at Harvard University discovered that sleep was divided into several distinct EEG levels, or stages, that unfold within an hour from drowsiness to shallow sleep to deep sleep. Because sleep recordings were stopped after an hour or so and not made continuously throughout the night, Berger and other scientists had failed to discover remarkable changes in brain activity that take place sometime after the appearance of deep sleep. Those observations, which were not made for nearly two more decades, revolutionized our understanding of sleep and its contribution to wakefulness.

In 1951 Eugene Aserinsky, a graduate student in the sleep laboratory at the University of Chicago, set out to study the slow eye movements that accompany sleep onset. He detected these rolling movements by observing the shifting bulges of the cornea (the transparent front coat of the eye) under sleeping infants' thin eyelids. In the course of his observations Aserinsky made a remarkable discovery. At various times during sleep the infants' eyes vigorously darted back and forth, up and down. These rapid eye movements appeared similar to those observed in the waking state.

Were the eye movements merely muscle twitches? Or were the infants scanning mental images in their sleep—were they dreaming? Did the sleeping brain suddenly become very active right in the middle of the night? How often and for how long did this activity occur?

In a fascinating series of now classic experiments Aserinsky, William Dement, and their mentor, Nathaniel Kleitman, began to unlock the mysteries of the night. They attached a pair of small recording electrodes to the faces of adult volunteers, near the outside corners of their eyes. With the help of an EEG machine to amplify signals from the eye muscles, any horizontal eye movements could easily be detected under the sleeper's closed eyelids. Another pair of electrodes was attached above and below one eye to measure up-and-down eye movements. (There was no need to attach electrodes above and below both eyes, as the eyes move together in binocular synchrony.)

At different times during the night the slow rolling movements seen at sleep onset were absent altogether or dramatically changed, to very active, rapid movements. Sleeping subjects were awakened during the periods of rapid eye movements (called REMs) and asked, "Was anything going through your mind?" The typical response was "Yes, I was having a dream." When subjects were asked the same question during non-rapid-eye-movement (NREM) periods, dreaming was reported much less frequently.

The Chicago team of pioneer sleep investigators had provided an objective index of when, and for how long, people experienced a state in which dreaming frequently occurred. The discovery of rapid eye movements with the accompanying dramatic changes in brain waves, heart rate, respiration, and genital arousal opened the floodgates for research on the sleeping brain and its effect on daytime functioning. Thousands of scientific studies began to fill the pages of newly established professional sleep journals. No longer could anyone regard sleep as just an inert, uniform, or passive brain state that served little purpose other than rest.

OBSERVING A TYPICAL NIGHT'S SLEEP

Many people think that soon after going to bed we slowly fall into deep sleep, remain there for some time, perhaps have an occasional dream, and then move gradually toward lighter sleep and awakening. This is far from what actually happens during sleep. Let's visit a sleep lab, hook you up, and trace the architecture of a typical good night's sleep.

Before bedtime a sleep technician temporarily glues small electrodes to your scalp in order to record brain waves. Other devices can be used to record eye movements, muscle tonus, body temperature, breathing, heart rate, hormone activity, and genital arousal. The bundle of long recording wires now attached to you is connected through a junction box on the bedroom wall to a polygraph (EEG) machine or computer in a control room. The machine will amplify the tiny electrical signals produced by your body and record traces on mile-long paper (for an eight-hour recording), or simply store the information in computer files for later playback and analysis. The technician turns off the bedroom light and wishes you a good night's sleep, which, believe it or not, is quite possible despite the array of wire leads emanating from your head, chest, and elsewhere. It's the technician, having to stay alert the entire night in the control room while you're sleeping peacefully, who will truly suffer. Such is the price of being a dedicated sleep specialist!

The Journey to Slow-Wave (Deep) Sleep

Lying awake in bed with your eyes scanning the ceiling, your fast (fifteen to twenty cycles per second), low-voltage (less than fifty microvolts) *beta* brain waves indicate wakefulness, as do the recordings of your eye movements (see Figure 3.1). Tired and ready for sleep, you soon shut your eyes and begin your night's slumber. Your brain waves become somewhat slower in frequency, higher in voltage, and more regular. These eight- to twelve-cycle-per-second *alpha* waves look like the teeth of a comb and delineate a relaxed, less tense, yet wakeful state.*

After several minutes in the alpha stage, your breathing rate begins to slow, as do your brain waves. You are now entering Stage 1 sleep, a transitional stage of light sleep marked by

---

*Beginning in the 1960s, some people used biofeedback devices to train themselves to enter and sustain this relaxed alpha state in order to reduce hypertension and headaches. The Israeli sleep researcher Peretz Lavie reported that although many studies clearly proved that control of alpha waves could be achieved, there was no solid proof that it significantly improved health. Nonetheless, it *is* relaxing, and that's worth something in today's stressful environment.

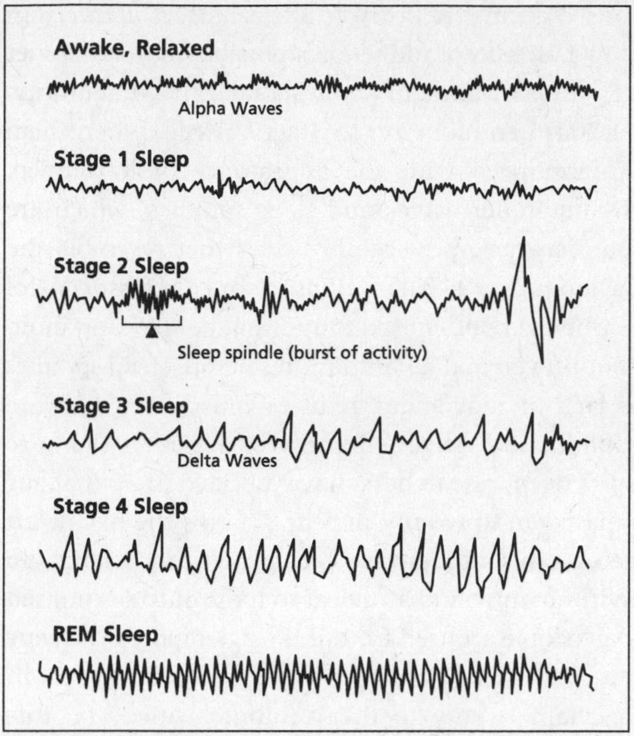

Figure 3.1 EEG waves of the awake and sleeping brain.

four- to eight-cycle-per-second, 50- to 100-microvolt *theta* brain waves.[2] In this twilight zone heart rate is lowered and stabilized, and breathing becomes shallow and regular. Sometimes it's hard for an observer to detect any breathing movement at all. This stage can last from ten seconds to ten minutes and is sometimes accompanied by fleeting visual imagery, so-called "hypnagogic hallucinations." Because the skeletal muscles suddenly relax, you might experience a sensation of falling, causing you to awake momentarily with a jerk (not referring to your bed partner!).

People aroused during theta-wave Stage 1 sleep often report they were only half asleep. Indeed, it's possible to maintain an awareness of your environment and respond somewhat quickly in this stage. You then move on to Stage 2 sleep, when theta waves are intermingled with the appearance of K-complex (single, high-amplitude) waves and sleep spindles, which are twelve-to-fourteen-cycle-per-second waves that resemble the spindle on a loom. Dr. J. Allan Hobson and Professor Wilse Webb of the University of Florida found that people stop moving in bed about seven and a half minutes before sleep spindles appear. The lack of movement reduces muscle tension and brain-stem stimulation via muscle nerves, thereby helping to induce sleep.[3] Sleep researchers have decided that Stage 2 sleep, which lasts ten to twenty minutes, marks the beginning of actual sleep. At this moment we become actively disengaged from our environment, blind and deaf to most outside stimulation. Nearly all people aroused from this stage report they were indeed asleep.

It's now perhaps twenty to thirty minutes since you first closed your eyes to seek the peace of sleep. You begin to enter Stage 3 sleep, a combination of theta and *delta* (very-low-frequency, high-voltage) brain waves. Soon, the theta waves disappear altogether. You have now arrived at Stage 4, the deepest phase of sleep. You have only delta waves, of .5 to 2 cycles per second, with an amplitude of 100 to 200 microvolts.[4] If you're awakened by an alarm or telephone call you'll feel mentally groggy for several minutes. You won't be able to make much sense. Your sound sleep has been abruptly interrupted. Young children are very difficult to arouse in Stage 4 delta sleep. You can literally stand them up and walk them around without disturbing their slumber at this time.

In delta sleep, muscle relaxation is complete, blood pressure drops, and pulse and respiration are slowed. Blood supply to the brain is minimal. We are at our most vulnerable time, all but free from the demands of our environment. It's difficult to be aroused from delta sleep. It's as close to hibernation as you get. So sleep tight, now is your chance.

Slow-Wave Sleep for Peak Daytime Performance

Slow-wave sleep is known for its restorative and growth-inducing properties, and plays a major role in maintaining our general health.

1. Restoration and Growth

- Blood supply to the muscles is increased in deep, delta sleep; this is when body recovery takes place. Deep sleep increases significantly on the nights after periods of sleep deprivation or vigorous exercise.[5] Studies of runners in a fifty-seven-mile ultramarathon race showed they experienced more Stages 3 and 4 deep sleep than usual in the two nights following their competition.[6]
- Body temperature is turned down in deep sleep, thereby conserving energy.
- Our metabolic activity is at its lowest in deep sleep, providing an opportunity for tissue growth and repair.
- The secretion of a growth hormone by the pituitary gland reaches its twenty-four-hour daily peak during Stages 3 and 4 deep sleep. The growth hormone stimulates growth and development, and also repair of the body's tissues. That's why uninterrupted deep sleep of significant duration is especially critical for children and adolescents. Less growth hormone is released as we age and older adults spend less time in deep sleep.

It is obvious that slow-wave sleep fulfills an essential need of the body to restore energy for future performance.

2. Immunity to Viral Infection

Natural immune-system modulators, such as interleukin and tumor necrosis factor, increase during slow-wave sleep.[7] Recent studies by Dr. Michael Irwin and his colleagues at the University of California–San Diego indicate that even a modest loss of sleep reduces the body's immune responses.[8] Simply put, if you're sleep-deprived, your resistance to viral infection is significantly lowered. Expect more colds and flu and respiratory tract infections. Luckily, the process is quickly reversible. Even if you didn't get enough sleep on a given night, the immune system will be back in action as soon as you repay that sleep loss.

Along with fever, sleepiness often accompanies illness. Your body seems to know that sleep helps fight infection. If you can manage to get plenty of deep sleep and meet your total sleep requirement every night, you are likely to see a remarkable improvement in your alertness, energy level, and general health. The details on how to do this are in Part III, "Preparing Your Mind for Peak Performance" (Chapters 5 through 8).

> O sleep, O gentle sleep. I thought gratefully, Nature's gentle nurse!
> —ELIZABETH KENNY
> (1886–1952)

If the sole purpose of sleep were bodily restoration, stimulation of growth, and immunity to viral infection, deep sleep would likely continue uninterrupted until you were fully recharged and ready for the next day's activities. However, things are not that simple. You haven't reached your final destination in the night's journey. You're not even close.

Ascending to REM Sleep

After thirty to forty minutes of quiescent delta sleep you begin retracing your steps back through Stages 3 and 2. It's now approximately 90 to 110 minutes after you fell asleep. You don't go back into Stage 1 "twilight" sleep. Rather, something dramatic happens. Your sympathetic nervous system is significantly more active now than it is in wakefulness or slow-wave sleep. Blood flow to the brain increases. Your pulse, respiration, and blood pressure increase and become somewhat irregular. Your body temperature rises. You have an erection or increased vaginal engorgement and lubrication. Your eyes, under closed eyelids, begin to dart back and forth, as if scanning the environment. Theta waves, no longer accompanied by K-complexes and sleep spindles, are intermingled with alpha waves, indicating a state similar to wakefulness. But you are not awake. You have arrived at Stage 1 REM (rapid-eye-movement) sleep.

This active stage is known as REM sleep because of its predominant eye movements. It is during this first REM period, which lasts for one to ten minutes, that you're most likely to experience your first dream of the night. Are your eyes visually tracking your dreams? Probably not: Congenitally blind people also experience rapid eye movements but do not dream "visually." It seems that a combination of distorted visual tracking of dreamed images and an involuntary contraction of muscles in the face owing to the activity of the trigeminal nerve may cause the movement. The root of this nerve is near a brain-stem region that emits bursts of electrical energy during REM sleep. The three branches of the nerve lead to the face and may affect the eyes.[9]

In REM sleep neuronal messages from the motor cortex of the brain are blocked at the brain stem. As a result, your muscles are completely relaxed and you are unable to move. Thus, REM sleep is characterized by an active brain, dreaming away, in what amounts to a "paralyzed" or motionless body. If we could move while we are dreaming we might injure ourselves and our bed partners. Unfortunately, in some older males, the area in the brain responsible for stopping motor cortex impulses is damaged and movement is possible during REM sleep. These men report acting out their dreams, sometimes causing serious harm to themselves and others (see Chapter 13, "Insomnia and Beyond," page 179).

During REM sleep males typically have penile erections and women experience clitoral erections, increased vaginal blood flow, and uterine contractions. This is typically unrelated to dream content or sexual activity or arousal and is quite normal, occurring in about 95 percent of REM cycles.[10] REM-period erections send blood into the penis, a process that brings oxygen and nutrients that help maintain healthy sexual functioning in the waking state. A physician who wants to check for possible psychological or physiological causes of male impotence can send his or her patient to a sleep lab for an all-night sleep recording. If the patient experiences REM-period erections, a physiological cause for impotence can probably be ruled out.

> **WHY ARE THERE PENILE ERECTIONS DURING REM SLEEP?**
>
> A final exam question in my university seminar on sleep is "What is the cause and purpose of penile erections during REM sleep?" The answer: Parasympathetic nervous system innervation and increased testosterone levels cause the physiological response; we don't know the purpose. However, one unprepared but creative student ventured: "I believe it serves as an antenna for receiving prophetic dreams." Full credit? You be the judge.

It should be noted that dreaming can occur in all stages of sleep, although dreams occur most frequently in REM sleep and are usually more vivid and emotional than dreams in other sleep stages.

REM Sleep for Peak Daytime Performance

REM sleep plays a major role in facilitating memory storage and retention, organization, and reorganization, as well as new learning and performance. Without the power of REM sleep, we would literally be lost, mentally.

1. Memory Storage and Retention

REM sleep is often referred to as "paradoxical" sleep because it's more like wakefulness than sleep in terms of vigorous brain-wave activity. The intensive firing of neurons spreading upward from the brain stem is thought to be responsible for aiding memory storage and retrieval as well as for reorganizing and categorizing information.

When we learn or experience something, certain neurons in our brain form specific connections with other neurons. This occurs through electrical stimulation of neuronal pathways and resultant changes in the neurons' protein structure. These chains of neurons, called neural networks or memory traces, are spread throughout the brain and are the repositories of all that we know. They are the filing cabinets in which every memory is stored, and from which every idea is drawn.

It is during REM sleep that much of the growth of specific neural connections to physically hold memories in the brain takes place. The evidence for this is compelling:

- Positron emission tomography (PET) scans of the brain show that metabolic activity associated with learning is significantly higher during REM sleep than during non-REM sleep or wakefulness (see Figure 3.2). Cerebral glucose metabolism is thought to indicate that REM sleep promotes memory.[11]

PET Scan of Awake, NREM, and REM

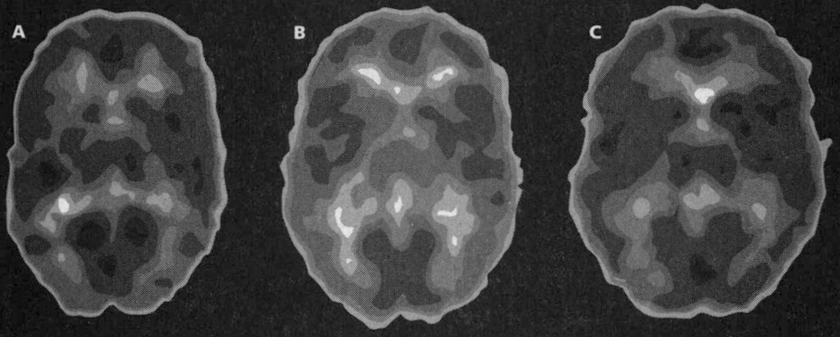

A   B   C

Above are PET scans of the brain (A) awake, (B) during NREM sleep, and (C) during REM sleep. Each shading of gray (actual PET scans are in color) corresponds to a different level of metabolism, with the black and white areas representing the highest and lowest levels, respectively, of activity. The scan during NREM sleep shows the low level of metabolic activity during this period. Reprinted with permission from *Life Sciences* 45 (1989), Elsevier Science, Inc.: M. S. Buchsbaum, J. C. Gillin, J. Wu, et al., "Regional Cerebral Glucose Metabolic Rate in Human Sleep Assessed by Positron Emission Tomography."

Figure 3.2. Positron emission tomography (PET) scans of the brain.

- There is more intense REM activity following periods of intensive learning. College students show an increase in REM sleep for several days after they study for final exams.[12]
- When sleep is disrupted, the brain's ability to transfer short-term memory into long-term memory is impaired.
- If skill training is intensive before sleep, there will be a significant increase in REM sleep, and individuals with the greatest increase in REM sleep perform best.[13]
- Performance studies show that there is dramatic improvement in memory retention after REM sleep, as opposed to non-REM sleep or an equivalent time awake. Conversely, people who are deprived of REM sleep find it difficult to retain material recently learned.[14] Thus, adequate REM sleep is particularly important for peak daytime performance that calls for memory retention and recall.[15]

Unfortunately (or in some instances, fortunately), we don't remember everything that we have ever experienced or thought. Brain cells and neurons decay over time, especially through disuse. If the neural connections in specific neural networks are not frequently stimulated they weaken and we may lose the information they contain.

During the day we make use of ideas and memories, thereby stimulating many neuronal pathways. If something is important we tend to consciously think about it, thereby strengthening the neural connections that hold the thought. But how do we retain infrequently used knowledge and memories? Luckily, we have REM sleep to help perform that necessary cognitive housekeeping task.

During REM sleep brain synapses are automatically activated.[16] There is intensive random firing of neuronal pathways that hold experiences and information. It is likely that dreams are the products of such stimulation, causing us to recall prior events, anticipate new ventures, or weave the familiar with the strange.

Some dreams may be meaningful, the result of stimulation of neural circuits holding ideas and memories of importance to our psyche. Other dreams are harder to analyze and might be meaningless. It's possible they are formed from unconnected ideas and memories stimulated by chance in the course of REM-sleep activation. No matter what the relevance of dreams, REM-sleep neuronal stimulation causes strengthening of memory circuits much as lifting weights causes strengthening of muscles. It is a "use it or lose it" proposition, and REM sleep helps us use it.

The beneficial effects of sleep for memory consolidation take many days to become evident, as measured by information-processing tasks. Sleep researchers believe that the process involves relatively slow-acting hormonal changes as well as more rapidly acting neurological ones.[17]

2. Memory Organization and Reorganization

We cannot learn new things during sleep. Sleep, particularly REM-sleep brain activity, provides an opportunity to file important memories of the previous day in long-term storage. Since it would be inefficient to store every thought in its own file—retrieval would be slow and difficult—it's likely that during REM sleep, ideas are organized into neural networks of associated ideas already in the brain, like a computer filing system using "folders." In this way, new learning can be efficiently connected to older information, as the brain replaces, modifies, or enhances memory as need be. And recall of associated memories is also efficient.

The reorganization of neural networks during REM sleep is perhaps responsible for those occasions when we solve problems in our dreams. Many artists, musicians, and scientists claim to have done some of their most creative thinking during their sleep, although such claims are often anecdotal, not scientific, in nature. The sewing machine, the periodic chart of the chemical elements, and the two characters Dr. Jekyll and Mr. Hyde were all said to have been conceived in dreams. Friedrich August Kekule von Stradonitz claimed he discovered the molecular structure of benzene, the benzene ring, in his dreams. In a lecture to colleagues he con-

> In sleep a human being is finally brought back to himself, having sloughed off everything else. Bound hard and fast, fettered with fatigue, he at last drifts toward the cavern of the unknown. Some men have returned therefrom with poems fully written, or equations solved.
>
> —GABRIELLE ROY (1909–), *THE CASHIER*

cluded, "Let us learn to dream, gentlemen, and then perhaps we shall learn the truth."[18]

Some writers and researchers suggest that novel ideas and solutions are possible in REM sleep because we have easy access to memories and emotions. We can give our thoughts and images free rein. We are not distracted by the environment. We are not constrained by reality.

Memory prioritization probably occurs during sleep.[19] While important memories are strengthened during REM sleep, trivial events are probably not stored, or are later discarded from long-term storage.[20] This permits us to make room for new information that we will absorb during our waking hours.

3. New Learning and Retention Through Replenishment of Neurotransmitters

Neurotransmitters are the chemical messengers in the brain that enable neurons to communicate with one another. Certain neurotransmitters, such as norepinephrine and serotonin, are thought to be crucial for new learning and retention. They are in limited supply and are depleted over the course of the day by the brain's continuous waking activities.

During REM sleep the brain cells containing norepinephrine and serotonin are inactive. The amount of neurotransmitter being used decreases substantially and the neurotransmitter supply is replenished for use the next day. These neurotrans-

IS IT POSSIBLE TO LEARN WHILE ASLEEP?
Storing, reorganizing, and "prioritizing" information during sleep may give us the sense of learning, such as waking up having solved a problem. However, you cannot learn new materials for the first time while you are asleep.[21] In order to acquire information you must be awake, so don't bother responding to those ads for pillow speakers.

mitters are essential for learning and memory, and preventing the supply from building back up as a result of lack of sleep or poor-quality sleep interferes with the ability to learn and remember.[22]

There is much more to REM sleep than the nightly entertainment of dreaming. In sum, adequate REM sleep is vital for memory storage and retention, memory organization and reorganization, and new learning. As noted before, the first REM period of the night lasts approximately nine minutes. This is hardly enough time to complete the critical REM sleep tasks necessary for peak daytime performance. Luckily, there is much more REM to come.

The Nightly Sleep Cycle

After the first REM period you head back into Stage 2, Stage 3, and Stage 4 sleep—then back again through Stage 3 and Stage 2 to more REM sleep. This cycle repeats itself every 90 to 110 minutes until you wake up (see Figure 3.3). Depending on the length of time you sleep, you will travel through four to five cycles before morning.

Note that there are some changes in the time spent in various stages as the night progresses. After the second cycle, Stages 3 and 4 sleep are minimal or absent, and Stage 2 dominates non-REM sleep. With each successive cycle the time spent in each REM period increases—from twenty to as much as sixty minutes.

If you sleep a full eight hours you might have as many as four or five REM periods each night, lasting a total of one and a half to two hours. That's enough time for significant dreaming, although you'll probably recall few if any of your dreams from a given night.

But remember that dreaming is not the only function, or the most important function, of REM sleep. REM sleep is absolutely

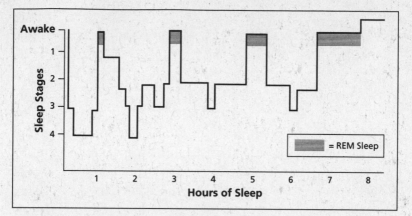

Figure 3.3. The architecture of a night's sleep (eight hours).

essential for preparing the mind for peak daytime performance. The fact that the brain needs REM sleep is dramatically illustrated by what is called the "REM rebound" effect. That is, when you follow several nights of too little sleep with a longer sleep, REM sleep will appear more frequently and for longer periods of time.

Whenever you have a short night of sleep you are eliminating the long REM sleep periods that come toward morning. As you will learn in the next chapter, such REM sleep loss produces serious daytime consequences in terms of your learning, thinking, memory, and performance. Nobody can afford REM sleep loss and be adequately prepared for success.

If it's now late at night don't stay up and read any further. You need all the REM sleep you can get to prepare your mind and body for tomorrow.

CHAPTER 4

SLEEP NEED AND PEAK PERFORMANCE

Everybody sleeps. But how much sleep do we need to feel fully alert, energetic, and primed for peak performance? And what are the specific consequences when we don't get all the sleep we need? First, we turn to a theoretical model that explains why we tend to fall asleep or stay awake at particular times during the night and day.

THE OPPONENT-PROCESS MODEL OF SLEEP AND WAKEFULNESS

Sleep experts Dr. William Dement and Dr. Dale Edgar have postulated that our brains possess two opponent processes that determine our tendency to fall asleep or remain awake: the homeostatic sleep drive and the clock-dependent alerting process.[1]

The Homeostatic Sleep Drive

Sleep is induced and maintained by our homeostatic sleep drive, a physiological process that strives to obtain the amount of sleep needed to provide for a stable level of daytime alertness. This process is active throughout the night and keeps us asleep. For most people, it takes at least eight hours of sleep to provide for sixteen hours of sustained wakefulness. Interestingly enough, the homeostatic sleep drive continues its work during the daytime. As our waking hours progress, the need to sleep is continuously building, although usually not so much as to overpower wakefulness. However, if we've slept too little during the previous night(s) and have a sleep debt, the tendency to fall asleep during the day will be significant, and we might indeed fall asleep at an inappropriate time. Conversely, an abundance of nocturnal sleep will reduce the tendency to be sleepy during the day.[2]

The Clock-Dependent Alerting Process

Wakefulness is induced and maintained by a clock-dependent alerting process, which is controlled by our biological clock. The biological clock is actually two tiny neural structures, called the suprachiasmatic nuclei, located in the center of the brain. The clock controls rhythms of alertness (but not sleep), body temperature, and hormone production. These rhythms are an intricate and orderly series of psychological and physiological changes that occur approximately every twenty-four hours and are called circadian rhythms (derived from the Latin *circa,* meaning "around," and *dies,* meaning "a day"). As we shall see, the rhythm of alertness can range from "foggy and slowed down" to "fully alert," depending on such factors as time of day, where we are, and what we happen to be doing at the moment.

Our biological clock, and hence the clock-dependent alerting process and its rhythms, is affected by exposure to light. Daylight signals the biological clock to stop the secretion of melatonin, a hormone that induces sleep, and to promote and consolidate a period of wakefulness. But what happens if you have no cues as to daylight and darkness? Will you always remain asleep, or at least not know when to wake up or stay up?

Biological-clock experiments have been carried out with subjects who volunteered to spend several weeks in a cave or a windowless room. The research provided two fascinating discoveries: (1) Our biological clock functions even in the absence of external time cues such as daylight and darkness, and (2) the clock-dependent alerting process runs, or oscillates, on a schedule close to twenty-five, not twenty-four, hours in length. So the subjects isolated from time cues tend to go to sleep one hour later each night.[3]

Living in a normal environment, you must resynchronize your internal twenty-five-hour clock every morning to adjust to the twenty-four-hour external day/night clock. This minor adjustment of one hour each day is easily tolerated by your body and usually goes unnoticed. But if you maintain vastly different bedtimes and rising times for the work week and on weekends, or if you are a shift worker frequently rotating between day and night schedules, or if you fly across multiple time zones, your biological clock needs to make a major adjustment to get in sync with your new schedule or local time environment. Otherwise your "alarm clock" for alertness will be buzzing at the wrong time in your sleep/wake schedule. You'll be wide awake when you want to sleep and sleepy when you want to be awake. (Information on how to cope with these vexatious problems is provided in Chapter 10, "Surviving as a Shift Worker," and Chapter 11, "Reducing Travel Fatigue.")

It's important to keep your biological clock in absolute synchrony with your daily routine. In this way the hours you spend in bed will correspond with the sleepy phase of your circadian rhythm and the hours you spend out of bed will correspond with the awake phase of your circadian rhythm. The only way to do this is to maintain a regular sleep schedule, going to bed at the same time every night and waking up at the same time every day, seven days a week, as described in Chapter 5, "The Golden Rules of Sleep."

Animals whose biological clocks have been surgically removed do not experience "any circadian periodicity in wake or sleep, and the daily episode of consolidated alert wakefulness is eliminated . . . but the ability to wake up and remain awake, at least for short periods, is not lost."[4] They will stay awake until the homeostatic sleep drive has increased enough to facilitate sleep.

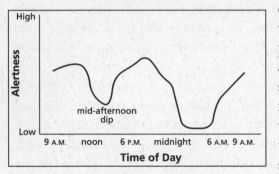

Figure 4.1. Alertness across twenty-four hours.

Humans without a biological clock to keep them alert for a consolidated daytime period would probably stay awake for one or two hours, then sleep for a few hours to repay the debt. It would be a very different existence, and probably not a very pleasant or productive one at that.

Independent of our sleep drive, there are fluctuations across the day in the strength of the alertness rhythm itself. Clock-dependent alerting appears to be moderately strong in mid-morning, quite weak in the early afternoon (the mid-afternoon dip, or trough), and fairly strong from late afternoon to mid-evening (see Figure 4.1). So we experience ups and downs in our alertness level over the course of the day, even if we've had adequate nocturnal sleep.

ASLEEP OR AWAKE?

Whether we are asleep or awake at any given time depends on the relative forces exerted between the two opponent processes, our homeostatic sleep drive and our clock-alerting (biological clock) process. They interact to produce our daily cycle of sleep and wakefulness (see Figure 4.2). During the daytime the drive for sleep is heavily countered by our clock-dependent alerting process, and for the most part we stay awake. As the evening progresses, our ability to think slows down because we need to fight the urge to sleep

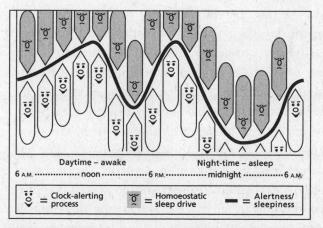

Figure 4.2. The opponent-process theory as postulated by Dale M. Edgar and William C. Dement.[5]

(the homeostatic sleep drive). Late in the evening the biological clock becomes inactive and sleep takes over.

Whenever we carry a sleep debt we are likely to feel sleepy in the daytime, as the drive to sleep will begin to overpower the clock-dependent alerting process. At particular times during the day, such as early to mid-afternoon, the rather low strength of the clock-dependent alerting process at that time will easily be overcome by a stronger drive to sleep, especially if we haven't met our sleep need at night. We look for an opportunity to nap.

The sleep drive is insidious—always there, ready to overtake the unsuspecting person who might be preoccupied with some engaging activity. You may not feel particularly sleepy at the moment, but the need to sleep is still present and can overwhelm you as soon as you stop doing anything stimulating. You are now at risk of falling asleep instantly, especially if you are carrying a large sleep debt. Let's hope you aren't driving a car on a monotonous highway . . .

THE SLEEP-DEPRIVED MAJORITY

You're probably getting the right amount of sleep if you feel alert all during the day, with no slump or fatigue until your regular bedtime. If you're not sleepy in sedentary situations, such as driving a car or sitting in a boring meeting during the mid-afternoon trough in alertness, you're among the alert minority and needn't read further. Give this book to a sleepy friend.

Many busy executives mistakenly assume they are good sleepers because they fall asleep immediately when they get into bed, or when they're sitting in an airplane. This is a sure indication of sleep deprivation. The well-rested person takes fifteen to twenty minutes to fall asleep. Think how ridiculous it would sound to brag about being a good eater because you devour meals the instant they are put in front of you. Such behavior would be indicative of food deprivation, not good nutrition practices. Likewise, if you fall asleep the instant your head hits the pillow, you are manifesting a sign of serious sleep deprivation.

By far the majority of us _are_ significantly sleep-deprived, yet remain ignorant of how much it affects our mood, performance, and behavior. We often surmise we are doing just fine. Why? Because we feel alert as long as we're engaged in vigorous, interesting, challenging, and stimulating tasks. But we readily excuse any drowsiness we feel after a heavy lunch, or a low dose of alcohol, or if we're in a warm room, or listening to a dull lecture, or attending a boring meeting. We mistakenly attribute our sleepiness to these "causes" and not to an underlying sleep debt. But guess what? None of these events _causes_ sleepiness. Such situations simply unmask the physiological sleepiness already in your body. In fact, if you've had adequate sleep, things like warm rooms and dull meetings will make you bored, mad, uncomfortable, and restless, but not sleepy.

Even though most of us think we are doing okay, we might be unknowingly carrying around years of accumulated sleep debt. We slowly habituate ourselves over time to a low level of alertness,

"NOW THAT WE'RE ALL UPRIGHT AND AWAKE AGAIN I'LL CONTINUE MY REPORT."

Copyright, _USA Today_.
Reprinted with permission.

thinking that how we feel now is normal. The truth is that most of us are functioning at a level far from optimal, far from the level of alertness that enables us to be energetic, wide awake, happy, creative, productive, motivated, and healthy human beings.

YOUR SLEEP DEBT BANK ACCOUNT

Each of us maintains a personal sleep bank account. We need enough sleep in that account to maintain the stable condition of sleep homeostasis, which will keep us alert during the day. Any sleep you get is a deposit or asset; any hour of wakefulness is a withdrawal, or debt. What does the average person's sleep account balance statement look like? It turns out that half the adult population is carrying a *substantial* sleep debt.[6]

As every hour you spend awake increases your sleep debt, you must continually make sleep deposits in your account. **Most people need to deposit at least eight hours of sleep in the account to cancel the sleep debt incurred by sixteen hours of continuous alertness.** "You need to obtain the amount of sleep each night that does not create a carry-over sleep debt," says Stanford's Dr. William Dement.[7] Otherwise you'll be in debt and drowsy day after day.

A sleep debt can build quickly, much like finance charges on an unpaid bank balance, if you're burning the candle at both ends.

For more than two years a Phoenix medical technician tried to get by on two to four hours of sleep a day. She worked nights at—ironically—a sleep disorders center so she could take care of her infant daughter during the day. She said she wasn't a good wife or a good mother, and wasn't good to herself. She reported the following symptoms: heart palpitations, being dizzy, afraid to drive, being crabby and having mood shifts from being up 20 hours a day.[8]

Contrary to popular belief, a sleep debt does not dissipate by itself over time, and it's cumulative. A one-hour sleep loss every night for an entire week is equivalent to having pulled one all-nighter! The only way to repay your sleep debt is to get more sleep. You can't repay years of sleep debt by one night of good sleep, any more than you can compensate for years of overeating by a one-day diet.

It takes some time to repay debts and establish a good schedule, but it *can* be done. In the next few chapters we will be telling you how to accomplish this critical task. People who take the time to find out what their sleep requirements are and then do the necessary things to get good sleep always exclaim after four to six weeks of their new sleep-wake routine: "I never before knew what it was like to be awake!" It can be that profound a difference. But too few people know of their sleepy states of existence, simply because most of us don't understand our own sleep need and the power of sleep in preparing the body and mind for peak performance. We don't value sleep. We fail to recognize the nearly linear relationship between sleep deprivation and performance. It's time for you to change that— to seriously heed a wake-up call!

SLEEP REQUIREMENTS FOR OPTIMAL PERFORMANCE

Just how much sleep is required to be in balance—to experience a total sense of well-being and be fully prepared for optimal performance? Researchers are now finding evidence that our natural need for sleep might be as much as ten hours per night. People in cultures that are free of the demands of a modern industrialized society typically sleep this much.[9] But who among the harried and the busy would be willing to consider such a luxurious schedule? Devoting even eight hours to sleep seems like an unobtainable goal. Yet the consequences for less-than-ideal sleep can be troublesome if not serious.

Ten hours of sleep is operationally defined as our need because that's what is often required for optimal performance. Timothy Roehrs and Thomas Roth at the Sleep Disorders Research Center of the Henry Ford Hospital in Detroit, Michigan, have demonstrated that alertness significantly increases when eight-hour sleepers who claim to be well rested get an additional two hours of sleep. Energy, vigilance, and the ability to effectively process information are all enhanced, as are critical thinking skills and creativity.

While most of us can operate satisfactorily on eight hours of sleep, we are simply not at our best. Furthermore, there exists no safety margin for occasions when we get less than that amount, which unfortunately occurs all too frequently. As soon as we lose as little as an hour's sleep, we are more prone to inattentiveness, mistakes, illness, and accidents.

According to Stanley Coren, professor of psychology at the University of British Columbia, in the four days after we lose one hour of sleep following the spring shift to daylight saving time, there is a 7 percent increase in accidental deaths compared to the week before and the week after—a pattern that is reversed in the fall when we gain one hour of sleep on a given night.[10]

WHY ARE WE LOSING SLEEP AND BUILDING DEBT?

Because we live in a twenty-four-hour society most of us are *chronically* sleep-deprived, chronically in debt. Work pressures and family and social obligations often lead to long hours of wakefulness and irregular sleep-wake schedules. There are simply not enough hours in the day. So where do we cheat? We cheat on our sleep. America is becoming a nation of walking zombies to an extent perhaps exceeded only by Japan.

The human body has limits as to what it can endure without rest, and, sadly, most of us are grossly exceeding those limits. Sleep deprivation is akin to depriving yourself of food.[11] Reducing your caloric intake by a constant amount each day has a cumulative effect. If you continue to burn more calories than you consume, you will lose weight. Keep this up and eventually your body begins to consume itself. Taken to an extreme, you will starve to death. Because sleep loss is cumulative, it can have similar devastating effects. People who go several days without sleep often experience such detrimental symptoms as hallucinations, delusions of persecution, slowed reflexes, impaired judgment, and feelings of hostility.

SLEEP TRIVIA
The official record for going without any sleep was set by Randy Gardner in 1965. He stayed awake for 264 hours and 12 minutes, then slept for 14 hours, 40 minutes.

In addition to the hurried and harried who are not obtaining adequate sleep, approximately 40 million Americans suffer from chronic sleep disorders. An additional 20 million to 30 million individuals have intermittent sleep-related problems.[12] If you fall into any of these sleep-deprivation categories, your chances of being productive, in a good mood, and healthy are severely reduced.

America is a nation at risk. We always seem to have "miles to go before we sleep."[13] According to Dr. William Dement, chairman of the National Commission on Sleep Disorders Research, "The national sleep debt is larger and more important than the national financial debt."[14] Are you a contributor to this deficit? If so, you're far from being the person you can be—you're hurting yourself and perhaps innocent others as well.

THE CONSEQUENCES OF SLEEP DEPRIVATION

Imagine sharing the road with someone driving in his sleep or being a passenger in a Boeing 747 on final approach, your fate in the hands of someone who is barely awake. Sleep deprivation not only affects those with minimal sleep, but also can have grave consequences for well-rested others.

"Rest. That's what I need is rest" was the Eastern Airlines captain James Reeves's comment to the control tower on a September 1974 morning, 30 minutes before crashing his airliner at low altitude, killing the crew and all 68 passengers.[15]

A five-month-old boy died of heat exhaustion after he was forgotten for nearly ten hours in the backseat of a car. Robert Gaito, a computer programmer, was supposed to drop his son at day care at 7:30 A.M., but forgot the baby was in the car and went to work. He didn't realize his mistake until 5:15 P.M., when his wife went to the sitter, learned the baby was not there, and called her husband. According to the doctor who performed the autopsy, the boy appeared to have struggled furiously against his seat belt and died of extreme heat exhaustion in the "enormously hot" car. Even after he had been dead for some time, the infant's temperature was 106° F. The father was described by a coworker as "a dedicated, driven employee who put in a lot of extra hours, and had probably overworked himself that week to the point of distraction. He was overtired, I guess."[16]

You are never exempt from the debilitating effects of insidious sleepiness. No amount of motivation or responsibility, even in risky or potentially dangerous situations, can override the powerful and inevitable consequences of extensive or cumulative sleep loss: impaired performance and unintended sleep.

Examples abound. Fred Smith, founder and chairman of the board, Federal Express, once said, "You don't realize how fatigued you can be. I went to sleep one time in the Marine Corps—walking—and I walked about a mile, as best I can recall, until I fell into a ditch."

James Rich, a private pilot flying a one-hour trip from Springfield, Kentucky, to Crossville, Tennessee, fell asleep after putting his plane on autopilot. He woke up six hours later over the Gulf of Mexico with an empty fuel tank. Rich had no flotation device aboard. He did not know how to swim. Luckily, he was rescued by the Coast Guard, but his $70,000 plane sank.

In 1989 the supertanker *Exxon Valdez* crashed into a reef in Prince William Sound, Alaska, and spilled 258,000 barrels of crude oil, resulting in extensive pollution, loss of wildlife, and a $2 billion cleanup bill. The mate at the helm was described as being too sleepy to perform his duties: "Severely sleep deprived, and apparently asleep on his feet, [he] failed to respond to simple, clear signals to turn the vessel back into the shipping lanes."[17]

Failure to repay an accumulated sleep debt can lead to bankruptcy and produce bizarre behavior, severe drowsiness, or unintended sleep seizures at inappropriate times. It really does happen.

A ninety-year-old man accused of killing his wife of sixty-two years told his doctor and investigators that he strangled the eighty-six-year-old woman because her persistent cough kept him awake at night. "You don't understand what I've been through. The last three weeks have been terrible and I haven't had any sleep in three nights."[18]

A sleepy computer operator working late at night ran the same program over and over for eight hours, at a cost of more than $100,000 to his employer.[19]

A waste-treatment plant operator at a big oil refinery dozed off in the middle of the night and inadvertently dumped thousands of gallons of chemicals into a nearby river.[20]

A pickup truck carrying twenty people veered off the freeway into a ditch near Barstow, California, killing twelve and injuring eight others in one of the worst single-vehicle accidents in the state's history. The cause: a driver who had fallen asleep.[21]

At 3:58 A.M. on April 13, 1984, two Burlington Northern freight trains collided head-on on the single main track at Wiggins, Colorado. Five train crew members were killed and seven locomotive units and twenty-six rail cars were destroyed. Total damage was estimated to be $3 million. The National Transportation Safety Board determined that the probable cause was that the engineer and other head-end crew members of one train fell asleep.[22]

Forgetting to extend flaps before takeoff, shutting down engines in midair, trying to land without the wheels down, landing at the wrong airport—even nodding off—aren't all that uncommon once exhaustion reaches a certain level. . . . One Delta Air Lines Boeing 727 crew, worn out from four straight days of grueling predawn flights, almost crashed into some office buildings. . . . The jet made a premature descent through heavy clouds to 400 feet before the startled crew realized they were still nearly thirteen miles from the runway. One commuter crew fell asleep at 16,000 feet just before the plane cruised into a violent Midwestern thunderstorm on autopilot and almost went into a stall.[23]

A former United Airlines copilot reported feeling tired over Denver on a delayed late-night coast-to-coast flight. He dozed off when a stewardess entered the cockpit to talk to the captain and the flight engineer, ostensibly to keep them awake. He woke up several minutes later to find that the captain, flight engineer, and stewardess had all fallen asleep.[24]

Those suffering from sleep disorders know all too well the cost of sleepiness:

Sleep apnea has ruined my life. I am now disabled and have lost a well-paying job. I cannot work because I keep falling asleep. . . . I cannot watch television, read or attend movies or plays. I am unable to drive my car. I have lost contact with most of my former friends. Who wants to be around a person who falls asleep during dinner or conversation?[25]

Although you might be able to avoid severe levels of sleep deprivation, even a moderate sleep debt can seriously affect the quality of your life. The research findings are quite clear: Sleep loss is devastating to performance.

SLEEP DEPRIVATION AND PERFORMANCE
Here's what happens when you don't get enough sleep:

* **DAYTIME DROWSINESS.** Inability to get through the day without a temporary loss in energy and alertness, usually in the mid-afternoon trough. Feelings of inattentiveness and grogginess, particularly when doing soporific tasks. Especially likely to occur after a heavy meal or a low dose of alcohol, while you are sitting in a warm room, listening to a boring lecture, or participating in a dull meeting. Again, these factors do not cause

sleepiness. They simply unmask the physiological sleepiness already in the body.

- **MICROSLEEPS.** Brief episodes of sleep, lasting a few seconds at a time, which produce inattention and can result in accidents, even death.
- **"SLEEP SEIZURES."** Unintended longer episodes of sleep that occur as rapidly as a seizure, without warning, in a severely sleep-deprived person.
- **MOOD SHIFTS, INCLUDING DEPRESSION, INCREASED IRRITABILITY, AND LOSS OF SENSE OF HUMOR.** Mood is one of the first things to be affected by sleep loss. With even minimal sleep loss, our threshold for containing anger is lowered. We can quickly lose friends, anger spouses, upset negotiations, and make enemies.
- **STRESS, ANXIETY, AND LOSS OF COPING SKILLS.** Overwhelming feelings of not being able to cope, even with simple problems or moderate work loads. Increase in worry, frustration, and nervousness. Inability to maintain perspective, or to relax under even moderate pressure.
- **LACK OF INTEREST IN SOCIALIZING WITH OTHERS.** Wanting to avoid group participation or interacting with others owing to fatigue. Desire to disengage from the outside world.
- **WEIGHT GAIN.** Consumption of beverages and foods high in sugar content used as an aid to staying awake when sleep-deprived. Some people attempt to reduce anxiety or boredom through eating.
- **FEELINGS OF BEING CHILLED.** Often the result of trying to stay awake very late at night, after the circadian rhythm ebbs and causes body temperature to plummet.
- **REDUCED IMMUNITY TO DISEASE AND VIRAL INFECTION.** The body's natural killer immune cells stop functioning as sleep deprivation increases.
- **FEELINGS OF LETHARGY.** Loss of motivation to maintain present tasks or pursue new endeavors.

- **REDUCED PRODUCTIVITY.** Reduction in cognitive functioning and reaction time, including the following:
  - Reduced ability to concentrate
  - Reduced ability to remember (especially short-term memory)
  - Reduced ability to handle complex tasks
  - Reduced ability to think logically
  - Reduced ability to assimilate and analyze new information
  - Reduced ability to think critically
  - Reduced decision-making skills
  - Reduced vocabulary and communication skills
  - Reduced creativity
  - Reduced motor skills and coordination
  - Reduced perceptual skills

Would you hire a person with the above traits to work for you? Next time you interview someone for a job, ask her how many hours of sleep she gets per night. If it's six or less, watch out. There's a good chance that the above list is characteristic of what you're about to buy into. It could also be true of someone who sleeps seven or eight hours. As we will see in the next chapter, there are individual differences in how much sleep a person requires to be fully alert all day long.

Often we are totally unaware of our own reduced capabilities because we become habituated to low levels of alertness. Many of us have been sleepy for such a long time that we don't know what it's like to feel wide awake. Only when we're exhausted and/or fall asleep at an inappropriate time are we seriously reminded of the body's fundamental need for sleep.

Do any of the above traits describe your behavior? You may not have fallen asleep at work, at the wheel of a car, or while driving a supertanker through a narrow channel. But you probably are one of the millions who are far from being able to perform optimally. Just how far gone are you?

SLEEP DEBT INDICATORS

How do you know if *your* sleep bank account is in arrears? One way is to visit a sleep lab and take a Multiple Sleep Latency Test (MSLT) developed by Dr. William Dement at Stanford University. The scientifically proven premise behind this test is that the more quickly you fall asleep at any time during the day when challenged to do so, the more sleep-deprived you are. It might seem easy for you to fall asleep anytime, whether or not you're tired. However, no amount of motivation, including monetary prizes, can make a fully alert person fall asleep on demand.

Here's how the test is conducted. You are hooked up to a polygraph machine that measures whether you are awake or asleep by recording brain waves, muscle tonus, respiration, and eye movement activity. Beginning at 8 A.M., for a 20-minute period you are put in a quiet, dark, and cool bedroom and challenged to go to sleep as quickly as you can. If you do go to sleep you are awakened instantly. This test is repeated every two hours throughout the day until 6 or 8 P.M.

If you don't fall asleep within any of the twenty-minute tests, you're said to be fully alert. For example, eight- to nine-year-old children who have not yet reached puberty usually get adequate sleep and rarely fall asleep on any of the test trials. On the other hand, exhausted high school and college students, as well as people with serious sleep disorders such as narcolepsy and sleep apnea, fall asleep after only three to five minutes during any test period (see Figure 4.3).

Most of us, especially busy executives and senior citizens, are at least moderately sleep-deprived and will fall asleep after five to fifteen minutes in one or more of the twenty-minute test

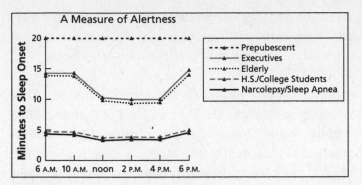

Minutes to Sleep Onset

20 ┄┄┄┄┄┄┄┄┄
15
10
5
0

6 A.M. 10 A.M. noon 2 P.M. 4 P.M. 6 P.M.

- ┄▲┄ Prepubescent
- ──▲── Executives
- ········▲········ Elderly
- ──▲── H.S./College Students
- ──▲── Narcolepsy/Sleep Apnea

Figure 4.3. Multiple Sleep Latency Test scores.

periods. This drop in alertness is most likely to happen between 2 P.M. and 4 P.M. in the mid-afternoon trough, when there is a noticeable dip in our clock-dependent circadian rhythm, even if we're fully rested. If you haven't been getting enough rest at night, this daytime drop in alertness is more severe and you'll probably be drowsy enough to fall asleep.

The Multiple Sleep Latency Test can detect as little as thirty minutes of nocturnal sleep loss from the previous week. It's a very accurate index of daytime alertness, far better than asking you if you feel sleepy. When you're preoccupied with stimulating work you may think you're alert even when you're actually quite sleep-deprived. Activity masks the sleepiness, but the tendency to sleep is there. If you were to stop work and do something rather monotonous and soporific, like take a Multiple Sleep Latency Test or drive on a freeway, the insidious sleepiness could reveal itself instantly and without warning cause a seizure of sleep.

A teenager who fell asleep at the wheel in the middle of the afternoon said: "Before I knew it I had drifted across the median and was on the other side of the road. A woman died and a little boy is still in a coma. He will probably have brain damage. I don't have a sleep disorder. If this can happen to me it can happen to so many people. It can happen to you. You are not invincible."[26]

In a study of young adults who declared themselves to be always alert during the day, Dr. Thomas Roth of Detroit's Henry Ford Hospital found that 34 percent of them were severely sleep-deprived as measured by the Multiple Sleep Latency Test. They thought they were alert, but they were not. They were disasters waiting to happen.

You don't have to go to a sleep lab and take the Multiple Sleep Latency Test to see if you're sleep-deprived. Remember your answers to Self-test B in Chapter 2? All the items, repeated here, are good indicators of probable sleep debt. If three or more of the following describe you, it's quite likely you need to get more sleep:

- **I need an alarm clock in order to wake up at the appropriate time.**
- **It's a struggle for me to get out of bed in the morning.**
- **I feel tired, irritable, and stressed out during the week.**
- **I have trouble concentrating.**
- **I have trouble remembering.**
- **I feel slow with critical thinking, problem solving, and being creative.**
- **I often fall asleep watching TV.**
- **I often fall asleep in boring meetings or lectures or in warm rooms.**
- **I often fall asleep after heavy meals or after a low dose of alcohol.**
- **I often fall asleep while relaxing after dinner.**
- **I often fall asleep within five minutes of getting into bed.**

- I often feel drowsy while driving.
- I often sleep extra hours on weekend mornings.
- I often need a nap to get through the day.
- I have dark circles around my eyes.

If the sleep debt indicators in the self-test describe your behavior, you need to make a change. You are probably experiencing some of the performance consequences noted above, but might be totally unaware of how far your situation is from the ideal.

THE NECESSITY FOR TAKING ACTION

One night with minimal sleep will not harm you, although you're likely to be less spontaneous, flexible, and original the next day. With a little extra effort and concentration you'll be able to perform routine tasks. If you lose another night of sleep, even performance of mundane chores will suffer.

Most of us need at least one more hour of sleep every night than we get. The consequences of gradual yet continuous sleep deprivation might not be readily apparent at first glance, but over time, the insidious sleepiness will begin to take its toll on your mood, your performance, and your health. Your education, your job, and your family and social life will be affected. Unintended sleep seizures, for which there is no warning, could be embarrassing, costly, and even fatal.

Getting a good night's sleep on a regular basis will determine how close you can come to optimal living during your waking hours. If your journey through the night is too short, fragmented, or disrupted and you're not getting enough sleep (and most of us are not), you've got to change your practices. There's an easy way to determine your personal sleep requirement and a variety of strategies to help you obtain a good night's sleep on a consistent basis. That's the subject of the next two chapters, so try to stay awake, or read them tomorrow when and if you're alert!

A DIGRESSION: A BIG YAWN

Yawns are slow, involuntary movements of the mouth that begin with a slow inspiration of breath and conclude with a briefer expiration.[25] The word "yawn" is actually derived from the Old English word *ganien*, meaning to gape or to open wide.

Facts about yawns

- Humans of all cultures, nations, and races yawn.
- Yawning has been observed in humans as early as the first three months after conception.
- Crocodiles, fish, birds, and snakes (just to name a few) yawn too!
- The average yawn lasts about six seconds.
- Yawns often recur at intervals of about one minute.
- Yawns feel good. In fact, on a scale of 1 (bad) to 10 (good), the average rating people give to yawns is an 8.5.

Are you tired, just bored, or sick?

- The idea that yawning is a response to an excess of carbon dioxide or a shortage of oxygen is completely false.
- An attack of a series of uncontrollable yawns is a very good indicator that it is time for bed. If you're getting enough sleep and are still a habitual yawner, you must be pretty bored!
- Oftentimes you yawn because you're stretching, but stretches are not always accompanied by yawns.
- Yawns can sometimes be viewed as an indication that there is something wrong inside the body. Frequent yawning can be one

of the symptoms of brain lesions and tumors, hemorrhages, motion sickness, opiate withdrawal, encephalitis, and chorea. However, psychotics rarely yawn.

Beware! Yawns are contagious . . .

- It's a proven fact that upon witnessing someone yawn, people are tempted to imitate the act.
- Contagious yawning is thought to be an ancient way of communicating bedtime to the rest of a group of people.
- Studies have shown that it is the whole yawning face, not just the gaping mouth, that stimulates a yawn response. So, covering your mouth during a yawn doesn't save anyone else from being infected.
- All in all, yawns are a normal and healthy function of the body. Try not to fight them. A stifled yawn is never as enjoyable as a completed one.
- Research indicates that even reading about yawns can trigger them. You may be experiencing this now!

Source: M. Carskadon, *Encyclopedia of Sleep and Dreaming* (New York: Macmillan, 1993).

PART III

PREPARING YOUR MIND FOR PEAK PERFORMANCE

Early to bed and early to rise
Makes a man healthy, wealthy and wise.

—BENJAMIN FRANKLIN,
Poor Richard's Almanac

SLEEP

CHAPTER 5

THE GOLDEN RULES
OF SLEEP

> **THE GOLDEN RULES OF SLEEP**
> 1. Get an adequate amount of sleep every night.
> 2. Establish a regular sleep schedule.
> 3. Get continuous sleep.
> 4. Make up for lost sleep.

OPTIMAL SLEEP FOR OPTIMAL LIVING

Do you want to be alert, dynamic, and full of energy all day long? Be in a good mood? Be productive, creative, and capable of making good decisions? Be able to express yourself well? Have good concentration and memory? Not be unduly susceptible to disease and viral infection? In general, do you want a chance at optimal living?

If so, you'll need to get optimal sleep, the amount and quality of sleep that allows you to function throughout the day without feeling drowsy.[1] Without optimal sleep you don't stand a chance of reaching your potential. With optimal sleep you'll be amazed at what you can do with your newfound life. **Optimal sleep is Power Sleep.**

Where do you start? By becoming familiar with and practicing the four golden rules outlined here. When do you start? Tonight!

THE GOLDEN RULES OF SLEEP

1. Get an Adequate Amount of Sleep Every Night.

Identify the amount of sleep you need to be fully alert all day long, and get that amount every night. It will dramatically change your mood and your ability to think critically and creatively. It will also improve your memory, perception, reaction time, and motor coordination (important for such activities as driving a car, or a golf ball). Perhaps for the first time you will know what it's like to be fully awake. Your improved performance will surprise you. And it won't take but a few weeks to see and feel some results.

The average adult sleeps between seven and eight hours. For a very few indi-

The Olympic gold medalist speed skater Dan Jansen says he needs ten hours of sleep when he's in training. Bonnie Blair, the holder of the most gold medals (five) won by an American woman in any sport, says she was getting barely more than six hours before she heard me talk about the importance of sleep in determining performance. I've been teasing her: "Imagine how many more gold medals you would win if only you got more sleep!"

viduals, six hours of sleep each night might be adequate. One or two people in one hundred can manage to get by with five hours. For a significant number of others it might take as many as nine or ten hours of sleep to function at full capacity and be wide awake all day.

Your optimal sleep requirement is largely determined by heredity. If both of your parents were short sleepers and were never drowsy during the day, you might be lucky enough to require less than average sleep. A very few people, called hypersomniacs, have been known to manage on just three or four hours. They have so much time on their hands they often work two full-time jobs and are impatient with those of us who "waste so much time in bed."

Realistically speaking, most of us aren't willing to balance our schedules in order to get the optimal ten hours of sleep mentioned in the last chapter. **At minimum most people absolutely need to obtain at least sixty to ninety minutes more sleep than they presently get.** In support of this rule of thumb, when volunteers in a study were allowed to sleep as long as they wished, they consistently slept an extra hour or more than usual. Dr. Thomas Roth, at Henry Ford Hospital's Sleep Disorders and Research Center in Detroit, found that sleeping one hour longer boosted a person's alertness by 25 percent! And that's just one of the innumerable benefits of getting more sleep.

Your Personal Sleep Quotient

It's time for you to get beyond the rule of thumb. Precisely determine your personal sleep quotient. Here's how:

- Start by selecting a bedtime when you are likely to be able to fall asleep easily. Settle on a time at least eight hours before you need

Drawing by Leo Cullum; © 1992 The New Yorker Magazine, Inc.

to get up. Maintain that bedtime for the next week and keep track of the time you arise. You might wake up too early for a few days if you've been conditioned to a short sleep schedule, but if you're sleep-deprived, that maladaptive conditioning will soon give way to longer sleep.

NOTE: If you need an alarm clock to wake you up, if you find it hard to get out of bed in the morning, or if you're tired during the day, you haven't slept enough.

- If you haven't been sleeping enough, don't change your rising time. Instead, go to bed thirty minutes earlier than usual for the next week. Add fifteen to thirty more minutes each week until you wake without an alarm clock and feel alert all day.

- When you establish your correct bedtime, you might try to cut fifteen minutes and see if that produces feelings of drowsiness the next day. Then you'll know for sure if you've identified your individual sleep requirement.

Your needs and sleep habits will change at various times during your life. In Chapter 12 you'll see that infants, adolescents, the elderly, and pregnant women are among those who are likely to encounter shifting sleep patterns that require special attention. In any case, you cannot get *too* much sleep; getting extra shut-eye never hurts. Sometimes we feel "groggy" when we sleep long hours. It's because we wake up in the midafternoon trough of alertness.

2. Establish a Regular Sleep Schedule.

Go to bed at the same time every night, and wake up without an alarm clock at the same time every morning, including weekends. That's 7 days a week, 365 days a year. Regularity is important for setting and stabilizing your internal sleep-wake biological clock. Within six weeks, the hours you spend in bed will begin to synchronize with the sleepy phase of your biological clock, and conversely the hours you spend out of bed will correspond to the time when you feel most alert and refreshed. Keeping a regular schedule will make you feel significantly more alert than sleeping for the same amount of time but at differing hours across the week and the weekend. Within a few weeks such regularity will actually reduce the total sleep time required for full daytime alertness.

British sleep researchers and a team at Harvard Medical School found that "if you alter your sleep schedule by even a few hours, your mood deteriorates. Shift workers may experience more anxiety and depression partly because they are out of synch with their biological clocks."[2]

Avoiding Sunday-Night Insomnia

Many people try to catch up on lost weekday sleep by sleeping in on the weekends. If you sleep late on Sunday, you won't be very tired at your regular bedtime that night. You'll finally fall asleep well after midnight. Within a few hours your alarm clock will jerk you back into consciousness and you'll crawl to work with the Monday-morning blahs. Your body will still be sleepy when it's time to be refreshed and alert. You've given yourself jet lag without the pleasure of leaving home. (We'll discuss jet lag in detail in Chapter 11, "Reducing Travel Fatigue.")

The brain does not have a different biological clock for weekdays and weekends. Changing your bedtime and/or rising time on the weekends will disturb your sleep rhythms. If you do stay up late on a given night, get up at your usual time the next morning. To make up for the lost sleep you might consider a nap (see Chapter 9, "The Nod to Midday Naps"). The important point is to not disturb your sleep-wake rhythms any more than necessary. It will only lead to insomnia.

Is There a Correct Time to Sleep?

In 1757 Benjamin Franklin gave us the epigram "Early to bed, early to rise makes a man healthy, wealthy and wise." It would be more accurate to say "Consistently to bed and consistently to rise . . ."[3] As long as you fulfill your sleep requirement without interruption, it doesn't really matter what time you go to bed or get up. But most of us have to get to work on a daytime schedule. Late-morning arrivals at the office aren't exactly appreciated—even if you're more fully rested and alert when you get there.

"Grandmother" psychology tells us that sleep before midnight is best, and that one hour of sleep before midnight is equal to two hours of sleep after midnight. Again, "It ain't necessarily so." While the first few hours of sleep are most restful in terms of deep (delta) sleep and the secretion of the growth hormone, it doesn't matter what the time on the clock is when such sleep occurs. Duration of sleep and regularity are what count.

One more point. You don't wake up "bright eyed and bushy tailed." You gradually become more and more alert, reaching a high point in the late morning and again in the early evening. So if you have to be at your very best first thing in the morning every day, you'd better plan on getting up a little earlier, having already enjoyed a sleep of optimal length.

Early-morning newscasters and talk-show hosts must get to bed around 9 P.M. and wake up at 5 A.M. for an 8 A.M. broadcast if they want to be close to optimally alert. Should they maintain that sleep-wake schedule on the weekends when they're not working? You bet. If they stay up Friday and Saturday nights to socialize and sleep in on Saturday and Sunday mornings, they'll have Sunday-night insomnia for sure. It's a hard life, but regularity is essential to success.

3. Get Continuous Sleep.

For sleep to be rejuvenating you should get your required amount of sleep in one continuous block. Disrupted nocturnal sleep is not restorative and will cause you to be drowsy during the day. Six hours of good, solid sleep is often more restorative than eight hours of poor, fragmented sleep. Sometimes sleep intrusions are difficult to avoid. Parents of newborns and males with prostate problems are all too familiar with the devastating daytime consequences of disturbed nocturnal sleep.

Limiting your time in bed to what you need, and no longer, will deepen sleep. Don't allow yourself to doze on and off for many hours. Many elderly people have medical conditions that make it difficult to enter or maintain deep sleep. As light sleepers they wake up frequently during the night. The resulting sleep deprivation causes them to doze off during the day, which in turn makes it difficult to fall asleep at night. This vicious cycle destroys the normal sleep-wake schedule and induces insomnia (see Chapter 12, "Avoiding Family Sleep Traps").

Keeping a regular bedtime schedule and avoiding long naps during the day will help everyone, young and old, have deeper and longer nocturnal sleep.

4. Make Up for Lost Sleep.

Because of our hectic twenty-four-hour society, it is likely that your sleep bank account will be in debt from time to time. Work deadlines, long trips, family obligations, and social events can lead to long days and short nights. An occasional late night won't do much damage to your alertness. But remember, reducing sleep by one hour for seven nights has the same effect as staying awake for twenty-four consecutive hours once a week. **Pay back your sleep debt in a timely fashion. Make up for any lost sleep as soon as possible.**

Pulling all-nighters to cram for exams is the students' desperate solution to poor scheduling and study habits. As noted in Chapters 3 and 4, the lack of sleep negates the chance for long-term retention of lessons learned, leaves students severely sleep-deprived, and lowers their immunity to infection. If you hear a lot of coughing and sneezing in a classroom it's a good bet that exam time has been close at hand. Always get a good night's sleep before any exam or other measure of your ability. According to the sleep expert Dr. William Dement, if you are in desperate need of more preparation time, consider four hours of sleep the absolute minimum, which is well below the optimal amount.

As you learned in Chapter 4, "Sleep Need and Peak Performance," sleep loss does not dissipate by itself over time. Furthermore, it is cumulative. If you lose several hours of sleep on a given night, you will become more and more sleepy in the ensuing days, even though you are again getting your "normal" sleep. Lost sleep must be repaid, although not hour for hour, if alertness is to be restored. If you don't remember this from Chapter 4 you're too sleep-deprived and should read the following points very thoroughly.

- You cannot replace lost sleep all at once. If you lose two nights of sleep you will not sleep for fourteen or sixteen additional hours on the third night. You may have significantly more slow-wave sleep for the next several nights, but you're likely to increase your total sleep time only by two to four hours. This is because your "sleep-wake" cycle depends on your sleep need and on your long-established internal timing mechanism: your biological clock, which is programmed to wake you up at a certain hour.
- When you sleep longer to catch up, try to do so by going to bed earlier than usual. Otherwise your normal waking time will be shifted. This is likely to make it difficult to get to sleep at the usual time the following night.
- You cannot make up for large sleep losses during the week by sleeping in on weekends any more than you can make up for lack of regular exercise and overeating during the week by working out and dieting only on the weekends.
- Sometimes a nap during the day will help you pay back your sleep debt. However, if you nap too long, or if you don't nap every day, you might interrupt your regular sleep-wake schedule and find that you do not sleep well that night (see Chapter 9, "The Nod to Midday Naps").

The important rule is to return to your regular sleep schedule as soon as possible. It might take as long as four to six weeks to repay years of indebtedness, but the resulting alertness, mental and physical functioning, and enjoyment of life will be more than worth the discipline.[4] The nice thing is that the cure for sleep loss and the resulting sleepiness is painless and pleasurable. If you want to be wide awake, creative, and dynamic all day long, all you have to do is **get more sleep!**

Now that you know the basic golden rules of sleep, it's time to explore various sleep strategies that will guarantee a better tomorrow. If you're not too tired, read on. Otherwise, turn off the lights and go to sleep. Come back tomorrow.

A DIGRESSION: EARLY BIRDS AND NIGHT OWLS

"Early to bed, early to rise." This maxim epitomizes the age-old attitude that morning people really are superior—at least a little healthier, wealthier, and wiser—than evening people. We now know that it doesn't matter when you sleep, as long as it's of restorative length.

"Early-morning types ('larks') leap out of bed at the crack of dawn, do their best work before noon, and become socially moribund in the evening. At the other end of the spectrum, there are the evening types ('owls') who cannot get their motors running in the morning but are still going strong way past midnight."[5]

Can Owls and Larks Live Together Happily Ever After?

More women than men are larks. More men than women are owls. Imagine trying to make a marriage work when you are a lark and your spouse is an owl. You complain that your partner is lazy and "sleeping the whole day away" while you are accused of being a "party pooper." You'd be lucky to spend any real quality time together; you are like ships passing in the night. Research has proved that when morning people and evening people marry, they spend less time in serious conversation and shared activities, including sex, than those who share similar sleep-wake cycles.[6]

So what can be done? The peak of the daily body temperature cycle tends to occur later in night owls than in people who are early birds. Even though some people are early birds or night owls because they have conditioned themselves to be, most are biologically inclined to follow their somewhat extreme sleep-wake schedules.

According to Mark Mahowald, M.D., "Most of us do not realize that some of the most important sleep characteristics of a given individual are genetically determined, and people have [no more] control over them than they do their height or their eye color."

Expecting your spouse to change an inherent sleep-wake schedule is "like marrying somebody who's five feet five inches tall and saying, 'Well, he'll be five feet eight over time.' "[7] The best solution would be to work around it, but don't expect things to change very much, at least until you're both senior citizens.

Here's something you can try if you want to change: Night owls should expose themselves to bright light soon after waking for several days (get out in the sun, but don't look directly at it). This will gradually cause the period of nocturnal sleep to occur earlier. "Morning larks" wishing to get up and go to bed later should conversely get bright light exposure before sleeping. Artificial sunboxes are made for this purpose.[8]

Lark or Owl for Life?

Our tendency to be owls and larks actually changes with age. As we get older, our body clock schedule shortens. Middle-aged people are generally morning people. The elderly usually become even more larkish, often retiring around 8 P.M., only to awaken hours before dawn.

Also, research has shown that perhaps the biggest transition in the sleep-wake cycle occurs as a result of the move from high school to college. While most high school kids are asleep by 12:30 A.M., the average college student doesn't crash until after 2 A.M.[9] This change is influenced more by social factors than by biological ones.

Some Words of Comfort for the Lonely Night Owl

If you're an extreme night owl who doesn't get sleepy until 3 or 4 A.M., don't worry! There is no "good" or "bad" sleep preference.[10] If you are a night owl you may feel as though you are somehow out of sync with our nine-to-five society. That's understandable.[11] As the lark lifestyle still predominates, owls are often considered by larks to be eccentric and sometimes lazy. But if the quiet evening hours work best for you and you don't have to be up for work at 7 A.M., go along with your inherent sleep-wake schedule. Many writers, artists, musicians, actors, and computer programmers choose to rise in the afternoon and work until very late at night.[12]

CHAPTER 6

TWENTY GREAT SLEEP STRATEGIES
How to Sleep Your Way to Success—Properly!

G ood sleep strategies are essential for anyone wishing to have restorative sleep and to feel totally energized day after day. There are literally hundreds of sleep strategies, some scientifically proven, some still anecdotal, and some just plain nonsense (those we'll not bother to mention here). What works well for one person often fails another, so it's up to you to experiment. The important thing is to not get too uptight about your sleep. A few nights of going with little sleep won't ruin your life, so relax, try things out, and see what works for you.

On any given night, one in four people has insomnia. Certainly you will not be immune to an occasional encounter with the frustrating and uncomfortable experience of having to do battle with the night. Insomnia is present when one or more of the following symptoms describes your sleeping pattern:

- You are unable to fall asleep when you want.
- Your sleep is fragmented and disturbed.
- You wake up too early in the morning.

What follows are specific suggestions to assure truly beneficial sleep, minimize the chance for insomnia, and guarantee daytime alertness.

1. Reduce Stress as Much as Possible.

Good sleep at night is strongly influenced by what happens during the day. For everyone except infants and the elderly, one of the most common reasons for insomnia is stress. In our twenty-four-hour society, stress is part of the fabric of everyday life. The Austrian endocrinologist Hans Selye noted that death is the only stress-free state. While we can never be entirely without stress, it can be managed.

Take control of your life. Focus on what's important, like family life. Put things into perspective before you become upset and/or angry. Keep in mind these two simple rules of stress management:

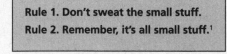

Rule 1. Don't sweat the small stuff.
Rule 2. Remember, it's all small stuff.[1]

According to the cardiologist Dr. Herbert Benson of the Harvard Medical School, meditative relaxation once or twice a day "can relieve the restlessness and tension that stand between you and a richer, fuller, healthier life."[2] Stress, anxiety, and fatigue are all relieved by what he terms the "relaxation response," which you can try right now. Here's the essence:

1. Sit quietly in a comfortable position.
2. Close your eyes and relax all your muscles, beginning with your feet and progressing to your face. Stay relaxed.
3. Breathe in easily through your nose. As you exhale, silently say a single word, like "one."
4. Continue for ten to twenty minutes. When finished, sit quietly with your eyes closed for a few minutes, then open your eyes for a few minutes before standing up.
5. Practice Dr. Benson's relaxation response once or twice daily, but not within two hours after a meal, since digestion may interfere with relaxation.[3]

A few minutes of peace and quiet every day, a positive mental attitude, involvement in sports and hobbies, lots of laughs, and the good company of loved ones will go a long way toward reducing tension. A tense person cannot sleep well. A relaxed person can. So relax.

2. Exercise to Stay Fit.

Exercise increases heart and lung fitness and reduces stress, anxiety, and insomnia. Exercise also raises your endorphin levels. Endorphins are naturally existing mood elevators produced by the brain in response to physical exercise. They reduce pain, relax muscles, suppress appetite, and produce feelings of general well-being. As a result, sleep will be deeper, more efficient, and more restful.

Effective exercise can include brisk walking, running, swimming, tennis, dancing, skiing, basketball, or aerobic workouts.

Consider exercise as a continuum. For sedentary people, even climbing the stairs to your office instead of taking the elevator would be an improvement. After checking with your doctor as to the best activities for you, make sure you schedule exercise on a regular basis. If you stop all exercise for as little as seventy-two hours, your fitness will begin to deteriorate. Moderate exercise even when you feel tired can make you feel more alert and help you sleep better that night.

Exercise elevates your core body temperature. An ensuing drop in body temperature at bedtime, five or six hours after a vigorous workout, induces drowsiness and deeper sleep. Athletes and other physically fit people have more delta (deep) sleep than do nonathletes.[4]

Do not engage in strenuous aerobic or physical muscular activity within three hours of bedtime. Physical exertion stimulates the release of adrenaline. You'll be too alert and might find it difficult to relax your body enough to induce sleep.

The best time to exercise is in the late afternoon or at noontime. Morning exercise has little effect on the quality of sleep that night. If you must exercise in the early morning, do not do so at the expense of needed sleep; make sure you get to bed in time to fulfill your sleep quotient. In the morning allow yourself enough time to raise your body temperature and become alert. Stretch before attempting a vigorous workout. It's easy to twist an ankle while jogging if you're still drowsy and not properly warmed up.

Do easy stretching before bedtime, but nothing strenuous. Move your arms in a Raggedy Ann, spaghetti-like fashion and try to relax your neck muscles by slowly rocking your head to the left and right.

An effective exercise regimen will strengthen your heart, lower your blood pressure, and reduce your blood pressure reaction to stress, as well as lower tension and anxiety. All of these will promote sounder sleep.

3. Keep Mentally Stimulated During the Day.

Boredom can cause loss of sleep. If you are physically and mentally active, you're less likely to get bored. Poor sleepers spend more time shopping, sitting around, and watching TV. Good sleepers spend time working, talking, doing chores, and pursuing hobbies. They are motivated and self-directed individuals, excited about life's opportunities.

Isolated people, depressed people, and those without a social support system of family and friends tend to be poor sleepers. Studies show involvement with others through projects and volunteer work helps such people reduce stress and anxiety by focusing on problems other than their own. Feel good about yourself, feel needed and loved, and you'll sleep better.

4. Eat a Proper Diet.

There is some controversy about the relationship between what you eat and how you sleep. In general, being healthy involves eating a proper diet. And healthy people sleep better.

- Eat vegetables and fruits, whole-grain cereals and breads, rice, pasta, fish, and poultry. Limit your intake of fat. Avoid fried foods.
- You should eat a basic healthy breakfast, a substantial lunch, and a light dinner.
- Eating proteins at dinner, such as fish, chicken, or certain vegetables will prevent hunger pangs at night. Do not eat a large or heavy meal within four or five hours of going to bed. While a substantial intake of food can make you feel drowsy initially, you'll probably toss and turn during the night. And you're likely

to gain weight by eating too late in the day to burn off the calories before they begin to do damage.

> According to Eunsook Koh, professor of clinical dietetics at the University of Oklahoma Health Sciences Center, "When you eat a big meal, the body dumps a large amount of insulin into your system, which helps to store fat. Add that to the inactivity of sleep, and the body will store much more fat than usual. You also are much more likely to have indigestion."[5]

- Stay away from foods that cause indigestion, gas, or heartburn. Pickles, garlic, and fatty or spicy foods are to be avoided at night. If you are sensitive to monosodium glutamate (MSG), you might induce insomnia if you eat a late-night pizza or Chinese food.
- If you're hungry at bedtime, a light snack high in carbohydrates and low in protein will settle your stomach and help you sleep.

> Carbohydrates with a high glycemic index—such as rice, potatoes, bread, and processed breakfast cereals—and fruits containing glucose or sugar help speed the amino acid tryptophan to the brain, where it is converted to serotonin, a sleep-inducing neurotransmitter. The process takes about forty-five minutes to one hour after the food is eaten. In contrast, proteins inhibit the transfer of tryptophan to the brain and help maintain alertness.[6] Cheese and crackers, cereal and milk, a glass of warm milk, sparkling water, Ovaltine, or a banana or apple can be helpful.

- Some herbs, such as valerian taken as a bedtime tea with honey, have sedative properties that can help put you to sleep and keep you asleep. Other noncaffeinated herbal teas include chamomile, lemon, and evening primrose.

- Certain vitamins and minerals, such as calcium and magnesium, are naturally occurring relaxants that can help you sleep better.[7] (Read *No More Sleepless Nights*, by Peter Hauri, director of the Mayo Clinic's Insomnia Program, and Shirley Linde, for an in-depth review of nutrition and sleep.)

5. Stop Smoking.

Aside from all its carcinogenic properties, nicotine stimulates brain-wave activity and increases blood pressure and heart rate. These factors all disturb your ability to get to sleep and remain asleep. If you have insomnia, stop smoking.

Nicotine is an even stronger stimulant than caffeine. Heavy smokers take longer to fall asleep, awaken more often, and spend less time in REM and deep NREM sleep. Because nicotine withdrawal can start two or three hours after the last puff, some smokers wake in the night craving a cigarette. When smokers break their nicotine habit, their sleep improves dramatically. Two-pack-a-day smokers who quit cut the time they lay awake in bed by almost half.[8]

6. Reduce Caffeine Intake.

Do not drink coffee, tea, or soft drinks that contain caffeine, such as colas, within six hours of your bedtime. As stimulants they delay sleep onset and disturb REM sleep, the period in which dreaming usually occurs.

Avoid chocolate (which also contains caffeine) and high-sugar foods near bedtime. Even that little piece of chocolate candy on your hotel pillow might disturb your rest.

7. Avoid Alcohol Near Bedtime.

You might be tempted to drink a glass of wine or something stronger to help you get to sleep. Having a nightcap is a com-

"How many times have I told you — NO COFFEE
AFTER SEPTEMBER!"
Courtesy Charles Almon

mon practice worldwide. But even widely shared customs are not necessarily a good thing.

In fact, you should **never** use alcohol to help yourself fall asleep! You might indeed fall asleep quickly, but your sleep will be disturbed; both NREM (deep, restorative) and REM (active, dreaming) sleep will be suppressed, and you will experience early-morning awakenings—often with a hangover. A drink before dinner, or a glass of wine with dinner probably won't make too much of a difference in your sleep. But avoid having any alcohol within three hours of bedtime if you expect to sleep well.

> You should avoid alcohol *at all times* when you're sleep-deprived. One drink of alcohol can make you stone cold drunk if you are carrying a large sleep debt. Sleep deprivation greatly magnifies the effects of alcohol, regardless of your physical build.

Drinking at bedtime can also start or aggravate sleep

apnea, a sleep disorder that causes you to stop breathing for up to ninety seconds several hundred times each night. Alcohol or sleeping pills might prevent the momentary arousals necessary to resume breathing, causing you to die in your sleep. (Read more about sleep apnea in Chapter 13, "Insomnia and Beyond.")

8. Take a Warm Bath Before Bed.

Just before going to bed, take a warm, soaking bath (around 100° F.) or relax in a Jacuzzi or hot tub if you're lucky enough to have one. This will send blood away from the brain to the skin surfaces and make you feel relaxed and drowsy. Your body temperature, raised by the warm water, will soon plummet if you have a moderately cool bedroom. This will initiate sleepiness and more deep (delta) sleep. (If you have a heart condition, high blood pressure, are subject to dizzy spells, or are pregnant, check with your doctor before subjecting yourself to very hot bathing.)

9. Maintain a Relaxing Atmosphere in the Bedroom.

You must condition yourself to associate the bedroom with pleasure and rest, not with stress and tension. Use the room only for sexual activity and sleep, not for arguing, watching exciting or violent television programs (even the late-night news can be upsetting), eating, working, or balancing checkbooks. Watching a comedy on TV, or listening to soft music might help as a tension reducer.

10. Establish a Bedtime Ritual.

When you were young your parents probably established a ritual that helped you get to sleep. They turned down the lights

and read to you until you began to nod off. Why not do something similar for yourself?

Engage in a nightly ritual of reading for pleasure just before turning off the lights. Find a good book, turn off the room lights, use a reading lamp that can be gradually dimmed, and take your mind off the day's worries by venturing into the author's thoughts. When you're fully relaxed or when drowsiness begins to lower your eyelids, you're ready to turn off the light.

11. Have Pleasurable Sexual Activity.

The one exception to exercising immediately before bedtime is sexual orgasm. Researchers have found that satisfying sexual activity (either through sexual relations or masturbation) can promote sleep onset and induce deep and restful sleep. Endorphins are released by sexual stimulation and can enhance the peaceful nature of sleep. However, if any sexual experience leads to dissatisfaction, anxiety, or concern about performance, it is likely to be more detrimental than helpful to a good night's sleep.

12. Consider Your Share.

Many couples find sound sleep is facilitated by the warmth and intimacy that comes with sharing a bed. But your situation may be different. Two people often do not rest as easily as one. Even sound sleepers move fifty to sixty times per night. If your bed partner has a sleep disorder you might be disturbed by kicking, flying elbows, tossing, turning, and snoring (see Chapter 13, "Insomnia and Beyond"). If that's the case, try a bigger bed, or

"I can't sleep."

Drawing by Ross; © 1993 The New Yorker Magazine, Inc.

put twin beds side by side. (See Chapter 7, "How to Create a Great Bedroom Environment," for more information on beds and mattresses.)

13. Avoid an Environment of Reigning Cats and Dogs.

While your pets might enjoy sharing your bed or your bedroom, their movements and noises during the night or early morning can disrupt your slumber. Unless you're greatly comforted by their presence and cannot bear to be without them for the night, dogs and cats can prove to be a hindrance rather than a help.[9] Speaking of bears, you might find a teddy bear to be comforting . . .

14. Clear Your Mind at Bedtime.

At bedtime avoid ruminating about problems and work. If your mind is too active, you'll have trouble falling asleep or you might wake up during the night or too early in the morning. Set aside your worries and calm your mind before going to sleep by writing each thought on a three-by-five index card on the nightstand. Jot down a potential solution, or a time during waking hours when you'll address the problem. That will transfer the problem from the brain to the paper. Leave it there for the night and get some sleep! You might prefer to use a pocket dictation recorder. If you wake up during the night and need to record a thought, you can do so without turning on a light.

15. Try Some Bedtime Relaxation Techniques.

If you're still having trouble falling asleep, try some relaxation techniques while lying in bed. Here are some tips from the Better Sleep Council:[10]

- **Progressive muscle relaxation (PMR):** Try tensing and then relaxing your muscles in groups, starting from the toes and slowly working up the body to the eye muscles and forehead. Squeeze tightly for five to ten seconds, then release and relax for fifteen to twenty seconds before moving upward to the next group.
- **Yoga:** Relax and inhale to the count of five, and raise your arms backward over your head until they touch the mattress. Make two fists and raise your buttocks. Tense and stretch every muscle, even your face. Then, arms still raised, let all the tension drain from your body.
- **Light a candle:** But only in your mind. Focus on the flame, and dismiss all thoughts that cause it to flicker. As it burns steadily, your mind becomes serene.
- **Mental imagery and fantasies:** Imagine yourself in a relaxing situation, such as lying on a tropical beach, strolling through

fields, floating through the air, or listening to soft music. Feel the warmth of the sun and the gentle breezes, hear the lapping sounds of the surf, and inhale the fragrances. Relax.

- **Deep breathing:** Take five deep breaths, and as you count each one, say to yourself, "I'm getting more relaxed, peaceful, and serene. I'm slowly falling asleep." Concentrate only on this message.
- **Mind games:** Imagine you're writing six-foot-high numerals on a large blackboard. Start at one hundred and count backward. You probably won't make it to fifty.
- **Counting sheep:** The oldest trick in the book really works. It distracts both sides of the brain with a soothing, repetitive activity; the right side sees the image, the left does the counting. You literally bore yourself to sleep. For those who can't visualize, we've seen a looped videotape of sheep jumping a fence. You play it on your bedroom TV; it repeats itself ad nauseum . . . you're sure to fall asleep.

If none of the above work and you are still wide awake, get out of bed until you feel drowsy again.

16. Avoid Trying Too Hard to Get to Sleep.

The number one piece of advice that sleep experts give to people with sleep problems is "Relax. You're trying too hard." Volunteers in a sleep study were offered twenty-five dollars if they could fall asleep quickly. Researchers found it took their subjects twice as long to fall asleep as it took a control group of volunteers who were not under such pressure to perform the pleasurable task quickly.

If you're not sleepy after a half hour in bed, or if you wake up in the middle of the night and can't get back to sleep, get up and leave the bedroom. Try to stay in the dark, or in dimly lit surroundings. Listen to soft music or do some light reading.

If you're too alert and feel like being productive, do some light housework—most of us get pretty tired doing that anytime. Soon enough you'll feel sleepy again. Go back to bed.

17. Limit Your Time in Bed.
Stress, depression, boredom, and partner pressure may get you going to bed earlier than you need to fulfill your sleep requirement. Older people, fearing a night of several awakenings and light sleep, often go to bed too early. This only adds to the problem of fragmented sleep. Go to bed only for that period of time you usually need for sleep, and sleep only until refreshed. Staying in bed too long will promote shallow and disturbed overall sleep.

18. Learn to Value Sleep.
If you want to enjoy life to its fullest, you must learn to value good sleep. You must do everything within your power to respect the needs of your body for rest. Leave the all-nighters and the late-night partying to others who don't care about their daytime alertness. You owe it to yourself not to settle for anything less than being wide awake and dynamic—at your best—all day long.

19. Use the Peak Performance Sleep Log.
Begin a new regimen immediately and track your progress. Don't waste another night or day. The time to start preparing for peak performance through the power of sleep is now. Begin by keeping track of your sleep habits and daytime alertness on a daily basis for the next six weeks. To help you do this systematically, you should complete the **Peak Performance Sleep Log** each morning at breakfast (see Figure 6.1). It will help you check on the regularity of your schedule, your sleep strategies, and your progress toward greater alertness through better

Peak Performance Sleep Log

Name: _____

Every morning at breakfast fill out the chart for the previous day and night.
For example, on Monday morning you should complete the "Sunday" column.

Nights:	Sunday	Monday	Tuesday	Wednesday	Thursday	Friday	Saturday
What time did you turn your lights out?							
What time did you get up this morning?							
How many total hours did you sleep?							
How many times did you wake up during the night?							
Rate the quality of your sleep last night. 1 = terrible to 5 = great							
Did you avoid taking a nap yesterday?	Yes ☐ No ☐	Yes ☐ No ☐	Yes ☐ No ☐	Yes ☐ No ☐	Yes ☐ No ☐	Yes ☐ No ☐	Yes ☐ No ☐
Did you avoid caffeine after 6 P.M.?	Yes ☐ No ☐	Yes ☐ No ☐	Yes ☐ No ☐	Yes ☐ No ☐	Yes ☐ No ☐	Yes ☐ No ☐	Yes ☐ No ☐
Did you avoid alcohol after 6 P.M.?	Yes ☐ No ☐	Yes ☐ No ☐	Yes ☐ No ☐	Yes ☐ No ☐	Yes ☐ No ☐	Yes ☐ No ☐	Yes ☐ No ☐
Did you do anything to reduce stress yesterday?	Yes ☐ No ☐	Yes ☐ No ☐	Yes ☐ No ☐	Yes ☐ No ☐	Yes ☐ No ☐	Yes ☐ No ☐	Yes ☐ No ☐
Did you avoid sleeping medications?	Yes ☐ No ☐	Yes ☐ No ☐	Yes ☐ No ☐	Yes ☐ No ☐	Yes ☐ No ☐	Yes ☐ No ☐	Yes ☐ No ☐
Was your bedroom quiet, dark, and cool?	Yes ☐ No ☐	Yes ☐ No ☐	Yes ☐ No ☐	Yes ☐ No ☐	Yes ☐ No ☐	Yes ☐ No ☐	Yes ☐ No ☐
Did you do anything to relax before falling asleep?	Yes ☐ No ☐	Yes ☐ No ☐	Yes ☐ No ☐	Yes ☐ No ☐	Yes ☐ No ☐	Yes ☐ No ☐	Yes ☐ No ☐
Did you eat a balanced diet yesterday?	Yes ☐ No ☐	Yes ☐ No ☐	Yes ☐ No ☐	Yes ☐ No ☐	Yes ☐ No ☐	Yes ☐ No ☐	Yes ☐ No ☐
Did you exercise yesterday?	Yes ☐ No ☐	Yes ☐ No ☐	Yes ☐ No ☐	Yes ☐ No ☐	Yes ☐ No ☐	Yes ☐ No ☐	Yes ☐ No ☐
How alert and energetic did you feel during the day? 1 = sleepy, tired to 5 = fully alert, energetic							

How are you doing? To be prepared for peak performance (5's in the last row):
1. You should be getting close to eight hours of sleep each night.
2. Your sleep and wake times should not change between weekdays and weekends.
3. Your sleep should be continuous, not fragmented.
4. Your sleep should be restful.
5. The answers to all the yes-or-no questions should be yes.

Figure 6.1. Peak Performance Sleep Log.

sleep. If you have followed the golden rules and sleep strategies in this book, by the end of six weeks you should see significant improvement in your alertness, mood, and ability to perform at an optimal level. Peak Performance Sleep Logs for the six-week period can be found in Appendix A.

20. If Necessary, Consult a Sleep Specialist.

Get on a regular sleep-wake schedule. Try one or more of the above techniques for a few weeks. See what works for you. It takes time. Be patient.

If this book doesn't put you to sleep (through helpful tips, or otherwise), it's time to seek more direct intervention from a health-care professional. You should always tell your physician if you are having problems with your sleep. His or her suggestions and prescriptions can be invaluable. If nothing seems to work after a month, see a sleep specialist. Get a referral from your doctor or call one of the accredited sleep disorders centers listed in Appendix D.

Get help immediately from a medically trained sleep specialist if you're waking up with shortness of breath and/or chest pain, or if you're falling asleep at inappropriate times, such as while driving a car or responding to a funny joke (see Chapter 13, "Insomnia and Beyond").

> Begin your new lifestyle tonight by going to bed fifteen minutes earlier than usual. If reading this has kept you up past your usual bedtime, forgive me. Start tomorrow night. Within just a few weeks of following the golden rules and strategies you'll join the ranks of the well rested. Your sleep bank account will be in balance, not in debt. You'll soon realize that you've never before known what it is like to be fully awake. The difference is that profound.

TWENTY GREAT SLEEP STRATEGIES

How to Sleep Your Way to Success—Properly!

1. Reduce stress as much as possible.
2. Exercise to stay fit.
3. Keep mentally stimulated during the day.
4. Eat a proper diet.
5. Stop smoking.
6. Reduce caffeine intake.
7. Avoid alcohol near bedtime.
8. Take a warm bath before bed.
9. Maintain a relaxing atmosphere in the bedroom.
10. Establish a bedtime ritual.
11. Have pleasurable sexual activity.
12. Consider your share.
13. Avoid an environment of reigning cats and dogs.
14. Clear your mind at bedtime.
15. Try some bedtime relaxation techniques.
16. Avoid trying too hard to get to sleep.
17. Limit your time in bed.
18. Learn to value sleep.
19. Use the **Peak Performance Sleep Log.**
20. If necessary, consult a sleep specialist.

> **BEGIN YOUR NEW LIFESTYLE TONIGHT**
>
> **If it's your bedtime**
>
> **Put down this book**
>
> **Take a warm bath**
>
> **Do easy stretching**
>
> **Crawl into bed**
>
> **Put out the light**
>
> **Try to relax**
>
> **Close eyes**
>
> **Sleep**
>
> **. . . Sweet REMs!**

CHAPTER 7

HOW TO CREATE
A GREAT BEDROOM
ENVIRONMENT

A cross the centuries and around the world people have slept well on everything from bare ground to hay, sheepskins, and waterbeds. Beyond the earth, astronauts have accommodated themselves to sleeping strapped vertically in sleeping bags under weightless conditions.

The bedroom is your refuge. Although people can get used to almost anything, the bedroom environment does make a difference in how well you sleep. For fully restorative sleep you must properly set the stage for the theater of the night.

SETTING THE STAGE

Most important, your bedroom must be quiet, dark, and cool.[1]

Noise

Make sure your bedroom is an easy place to fall asleep and stay asleep. Keep your bedroom quiet. Individuals differ in their sensitivity to noise. Any sound that exceeds seventy decibels can stimulate your nervous system, send signals to the rest of your body, and keep you awake. Dripping faucets, noisy radiators, ambulance, fire and police sirens, barking dogs, loud stereos, and late-night revelers are common rest inhibitors.

> **SLEEP TRIVIA**
> Winston Churchill had twin beds. When he couldn't fall asleep in one, he would move to the other. "He is known to have worked from his bed during World War II."[2]

Sounds should be low level and consistent, or eliminated altogether. Abrupt, intense, and intrusive sounds can disturb your sleep. A sudden noise will elevate your blood pressure. Changes in sound intensity will be more disturbing than constant sounds. If the sound increases in intensity, your pupils will dilate and your heartbeat will quicken. Perhaps it's good that an important signal like a baby crying for attention can wake you easily. But it won't help your sleep.

You might be able to adapt to certain familiar noises over time, such as the ticking of the grandfather clock, or traffic noise if you live in a city or near a highway. It takes about a week for someone from a rural area to adapt to city traffic noises. Conversely, the absence of familiar noises can also disrupt sleep. City dwellers often find it difficult to fall asleep when they vacation in the quiet countryside. It's too quiet! People who live near airports often say they sleep through the roar of engines overhead, but researchers have found that the noise still disturbs sleep. If you can't avoid noise from planes, traffic, neighbors, or even other family members, try earplugs, or a fan or air conditioner in the summertime. If music helps you sleep, keep a radio near your bed.

Mail-order companies love to promote fancy gadgets that play the sound of the surf or crickets in the forest to help put you to sleep. If you want to spend extra money and the electronic sandpersons seem to help, that's fine. Some people also find white-noise generators helpful for blocking out sounds that are keeping them awake. A cheaper alternative is to set the tuner of your FM radio between any two stations. The pseudo white noise you'll hear will do wonders to mask unwanted sounds and promote sleep.

Light Level

Too much light can contribute to sleeplessness. Using dark fabric to block windows or the rim of a door can significantly reduce the amount of light leaking into the bedroom from hallways, streets, and early-morning sunlight. Eyeshades may also do the trick.

Temperature

The ideal temperature for sleeping is about 65° F. If it gets too warm or too cold your total sleep time may be reduced. A hot bedroom, or too many blankets and bedclothes, may disturb sleep and even induce nightmares; you're more likely to wake up in a sweat.

Many parents bake their young children at night, although with only the best of intentions. It's no wonder the kids wake up in the middle of the night with bad dreams. Don't freeze your kids, either; a cold room could be just as disruptive if it causes them to wake up shivering during the night.

Humidity Level

Some people prefer fresh air, but it's not a prerequisite for a good night's sleep. An ideal relative humidity level for the bedroom is between 60 and 70 percent. You may want to buy a humidifier or dehumidifier, depending on your needs. An added advantage of having a humidity-control device is that its constant hum may drown out noises that could otherwise be disruptive.

Security

Do you feel safe and secure at night? An extra lock on the front door might make all the difference in how secure you feel at night; so too might another smoke alarm. As part of your nightly bedtime ritual, check the door locks and close windows before you head for the bedroom. You'll sleep easier. And don't forget to keep fresh batteries in your smoke detector.

Clocks

Hide illuminated clocks from view to avoid clock-watching during the night, which can lead to anxiety over sleeplessness. If you have established a regular sleep schedule you shouldn't need an alarm to wake you up, but set one anyway to avoid oversleeping—just in case you're more sleep-deprived than you realize!

Colors, Decor, and Cleanliness

For a restful bedroom, select colors that you associate with a feeling of peace and ease. For example, many people choose blues or greens because it reminds them of the ocean or of being outdoors in a park. Keep the bedroom clean and free from clutter. Piles of clothes, reports, and bills induce feelings of stress. Paintings of pastoral scenes, photographs from enjoyed vacations, and portraits of family members can create a peaceful ambience for rest.

Nightclothes

Are your nightclothes scratchy or otherwise uncomfortable? Are they too light or heavy for the season? Loose-fitting soft garments that breathe are preferable.

Bedsheets

The key for bedsheets is clean, cool, and comfortably soft. Popular bedsheet materials are cotton, linen, and polyester fabrics. "The quality of the fabric is determined by the number of threads woven per inch—the higher the thread count, the better the material."[3]

> **SLEEP TRIVIA**
> Charles Dickens believed that his body position and the location of his bed were all-important in overcoming insomnia. He checked to be sure that his bed was pointing due north. Then he would place himself exactly in the center of the bed, measuring the distances to both edges with his outstretched arms as he reclined.[4]

Cotton is the most popular fiber for bedsheets, used either alone or paired with linen, silk, or polyester. Cotton sheets are absorbent, making them practical for any climate, and long-lasting, becoming even more comfortable over time.[5] Silk sheets are also durable, but tend to be more costly. They have a luxurious feel and keep you warm in winter.[6]

For superior quality, you might try linen sheets. The "healthiest material for sleeping on,"[7] linen does not become soiled or absorb moisture as easily as cotton. Linen is ideally suited for use during the summer months or in hot climates because of its light weight and cooling effect. Though linen sheets are the most expensive, they are a good long-term investment; two sets of fine linen sheets can easily last twenty years.[8]

Consider the color of your sheets. For soothing sleep times, you may want to look for white or pastels, or soft designs like florals, paisleys, or light-colored striped sheets.[10]

PILLOW TALK

There are good pillows and bad pillows. A good pillow, essential to long-term health, gives you comfort and automatically puts you in a healthy sleep posture. Orthopedic doctors and chiropractors agree that nearly all neck and back problems are worsened, if not caused, by improper sleeping habits and bad or worn-out pillows. The Better Sleep Council makes the following points about pillows:[11]

- Sleeping on your stomach or with your head elevated can give you aches and pains. To give your body the proper rest it needs, and to ensure the health of your cervical spine (neck), medical experts recommend only two sleeping positions: Sleep on your side, with the spine straight; or on your back, maintaining the primary curvature of the cervical spine. Both these positions keep your cervical spine in natural alignment and prevent "subluxation" or misalignment of the neck. When you correctly support your neck, you help prevent neckaches, backaches, pinched nerves, and general health problems caused by wear on the vertebrae (back) and spinal column.

  Sounds good, doesn't it? But how do you suddenly give up unhealthy sleeping postures and acquire better habits? With the right kind of pillow.

- A good pillow gives you support in just the right places, whatever your sleeping position. For example, if you're a side sleeper, you need a pillow that supports both your head and neck and keeps them in perfect alignment. If you're a back sleeper, the support should be mostly under your neck, with your head almost touching the mattress. The amount and type of support you need depends on your size, weight, shape, sleep position, and the density of your mattress.

- For the best night's sleep, look for pillows that let you fluff and squish the pillow to fit your unique contours, shape, and sleeping posture. Your pillow should "fit," just like shoes. Buying a pillow that offers maximum adjustability assures you a comfortably healthy fit.

- Natural-fill pillows, such as down or feather pillows, give you the most comfort, long-term performance, and adjustability of any pillow. They gently support your head, providing tactile softness, eliminating pressure points, and increasing facial circulation to reduce face squashing, or "sleep wrinkles." Down pillows offer the optimum in comfort, luxury, and performance. They'll outlast any synthetic fiber, and therefore make a good long-term investment.

- Unfortunately most people hang on to a pillow long after it has lost its healthy sleeping power. Just because a pillow isn't falling apart doesn't mean it can still give you a healthy rest.

TEST YOUR PILLOW

Examine the covering fabric: Is it stained, disintegrating, or leaking filling? If so, you probably need a new pillow.

To test the support of down and feather pillows, lay the fluffed pillow on a flat, hard surface. Fold your pillow in half and squeeze out the air. (Fold a king-size pillow in thirds.) Release the pillow. A pillow with support will unfold and return to its original shape. A broken pillow will remain folded.

To test the support of polyester pillows, fluff and fold as above. Place a ten-ounce weight on the pillow (a medium-size tennis shoe is perfect). A pillow with support will throw the shoe and unfold itself. A broken pillow will remain folded.[12]

- The comfort lifetime of a pillow is the length of time you can expect it to retain its original support and adjustability. Over time, all pillows lose their fluffiness. Some people sleep aggressively, taking out frustrations on their pillows by tossing and turning. Others sleep gently, extending the life of their pillows. Assuming normal use and care, the comfort lifetime of feather pillows is eight to ten years, down pillows is five to ten years, and polyester is six months to two years.[13]

- You may have to spend as much as $70 for a good pillow. Think that's expensive? Compare your investment to that for a pair of good shoes. A good pillow will last you eight hours a night, 365 days a year, for up to ten years. That's 292,000 hours! If you're lucky, a pair of shoes will last one twentieth that time.

HOW TO SELECT THE RIGHT MATTRESS

How do you determine if your mattress needs replacing? The Better Sleep Council has an ABCs checklist to follow[14]:

A. **Age.** Studies have shown that a mattress and foundation provide optimum service for about eight to ten years of nightly use. After about one decade, a mattress no longer offers the comfort and support you need for a good night's sleep.

B. **Beauty.** The next time you change your sheets, take a critical look at your mattress. Would you be embarrassed to show your mattress without covers to your neighbors? Look for soils, stains, or tears in the cover, and uneven surfaces and sagging spots around the edges or where you usually lie. A mattress's poor appearance is generally indicative of its poor performance.

C. **Comfort.** Lie down and concentrate on the feel of your mattress. This might not be as easy as it seems. When you sleep on the same bed each night, you become desensitized to its dwindling comfort and support—the way you become used to an old pair of sneakers that no longer support your feet. If you find it difficult to gauge your mattress's comfort, visit a local bedding or department store to see how a new, good-quality bed feels. You'll discover that advancements in bedding technology have made today's beds more comfortable and supportive than new beds were a decade ago.

Did your mattress fail one or more of the ABCs? If so, it's probably affecting the quality of your sleep. Throw out your old mattress and invest in a new one.

A new Mercedes wouldn't perform like a Mercedes if it had old Chevy shock absorbers on it—and a new mattress shouldn't be put on top of an old foundation. Yet many people do just that. A mattress and foundation, or boxspring, are engineered to work together as a sleep set. Like a giant shock absorber, your foundation takes a lot of nightly wear and tear and contributes to the bed's overall comfort and ability to support your body. An old foundation will greatly reduce the life of your new mattress.

> **SLEEP TRIVIA**
> In 1969 the fakir Silki claimed to have spent 111 days on a bed of nails in São Paulo, Brazil.[17]

SELECTING A MATTRESS

If you have determined that you need to replace your mattress, how do you choose a new one? The Better Sleep Council offers some tips on how to "test-drive a bed" when you go to the store.[16]

1. Pretend the Mattress Is in Your Bedroom.

When you sit behind the wheel of that shiny new car in the showroom, you know how great it's going to feel when it belongs to you. So don't make the common mistake of selecting a bed without lying on it first. Simply sitting on a mattress or pushing on its surface with your hand won't tell you how it's going to feel to your weary body at bedtime. Wear comfortable clothes, and shoes that you can slip off easily. Ask the salesperson to give you some time, lie down in your normal sleeping position, close your eyes, and shut out the world around you. Don't let feelings of self-consciousness intrude on your relaxation. Give yourself a few minutes on each bed before moving on to the next one.

Trust your instincts. "Test-drive" any mattress you are considering. Don't get bogged down by product labels. One manufacturer's "firm" may be another's "extrafirm." Also, model names may differ from store to store, making it even more important to lie down and try each of the mattresses. Select a reputable retailer and let your instincts guide you in gauging each bed's comfort and support. Look for a mattress and foundation set that gently supports your body at all points. If there's too little support, you can develop back pain. A too-rigid mattress can create uncomfortable pressure after you have been lying in one position for several minutes. It will take you three to five days to become accustomed to a switch from a soft to a hard mattress.

SLEEP TRIVIA
The late eighteenth century saw the advent of cast-iron bedsteads and cotton mattresses. This made sleeping spaces less attractive to bugs, hitherto an accepted problem. The expression "Sleep tight, don't let the bedbugs bite" has a basis in fact.[18]

2. Look "Under the Hood."

As with your car, what's inside the mattress does count. Your retailer may have a cutaway model that shows you what is underneath the ticking; if not, ask about the mattress's construction. A king-size innerspring mattress should have more than 450 coils; a queen, more than 375; a double or full, more than 300. A good foam mattress will have a minimum density of 2.0 pounds per cubic foot; the higher the number, the better the foam. If you prefer the feel of flotation that comes from a waterbed, check to make sure the vinyl is at least 20 millimeters or thicker for good durability. Compliance with California's waterbed regulations is a good measure of performance, too.

On the outside, look for signs of quality: fine tailoring, superior fabrics, and a surface that looks and feels plush. Today's better-quality mattresses feature extra layers of luxurious cushioning for comfort. A word about the warranty: The warranty is your insurance against product defects, but it's not meant to tell you how long to keep your new mattress and foundation. A sleep set may be warranted for and may still be usable after fifteen to twenty years, but it probably will no longer be providing the optimum comfort and support you need for a good night's sleep.

3. Go for the Rolls-Royce.

You may have your eye on that sexy two-seater Italian sports car, but when it comes to beds, bigger is always better for sleeping comfort. It is recommended that you choose a bed about six inches longer than the sleeper. If you sleep with a partner, don't settle for anything smaller than a queen; you may be even happier with a king-size bed. Sleep studies show that we all turn between forty and sixty times a night, and that we require room to move about freely so that we don't awaken.

4. How Much Should You Spend?

The market offers a wide range of prices, from a few hundred dollars for a budget-priced twin-bed mattress and foundation set to more than $2,000 for a Rolls-Royce, ultra-premium king set. The best answer is: Buy the best you can afford. Your bed is an investment that should last longer than your car, and you'll be spending much more time in bed than behind the wheel in the next decade. How comfortable you are in bed affects the quality of your sleep—and ultimately the quality of your life.[20]

The bottom line in creating a great bedroom environment is to minimize discomfort and distraction. Arrange for whatever will assure you of a good night's sleep. **Remember: quiet, dark, and cool . . .**

A DIGRESSION: BEDROOMS IN SPACE
by Astronaut Gerald Carr, Commander, Skylab 4

For years I have described my eighty-four-day sleeping quarters on Skylab 4 as being the same size as a telephone booth. Actually, my room was a bit larger than that, but not by much. The shape was trapezoidal. One wall was floor to ceiling (about seven feet) drawers and lockers, one was the door which was of fabric (quilted beta cloth) that was fairly sound-proof, and the other two were aluminum. The floor was aluminum also with an air-conditioning register blowing upward, and the ceiling was a baffled grid that kept light out and permitted air flow out of the quarters.

My bed consisted of a sleeping bag laced to a frame (like a sailor's bunk), which was in turn fastened to the aluminum wall opposite the locker wall. Alan Bean, the commander of Skylab 3, was a fresh-air freak, so he inverted his bag on the wall so that he slept head-down with his face near the air inlet. When one is weightless and the lights are turned off, one is just as comfortable in either orientation. The sleeping bag itself had a net for the first cover layer and a fabric outer cover that could be unzipped and restrained open if the occupant wanted to be cooler. Across the bag at chest level and thigh level were eight-inch-wide adjustable elastic "cummerbunds" attached to the frame, which were used to push my body against the wall. The body pressure gave me the secure feeling when the lights were out that I was in a regular bed. My pillow was a soft foam block with a hair net attached above it, and when I was ready to go to sleep I merely hooked the net under my nose and ears, and it kept my head from lolling around as I slept. I tried sleeping free-floating in the largest compartment of the lab, but I needed my cummerbunds. I would awake with a start, feeling like I was falling out of my Earth bed.

As for sleep itself, I slept well. My norm is about eight hours, but in orbit it became seven. I think that it was probably due to the fact that I did not go to bed with the muscle fatigue that one has in one G on the Earth. In weightlessness one is a butterfly. You can go anywhere, and the stress on the skeleton and musculature is negligible. Only the brain gets a rough workout. We lived on Houston time, and sleep time was designated at 10:00 P.M. until 6:00 A.M. We saw fifteen sunrises and fifteen sunsets every day. The three science pilots (SPTs), Joe Kerwin, Owen Garriott, and Ed Gibson, wore electroencephalograph skullcaps to bed on occasion so that NASA could record their brain waves as they slept. I'm not sure that anything unusual was observed in those experiments.

On the fiftieth day of Skylab 4, I recorded in my diary that, to my surprise, I remembered dreaming my first weightless dream. Instead of moving from one place to another by walking, I did it by floating as we did on Skylab. It was remarkable to me that my subconscious wouldn't let go of its Earth roots for the first forty-nine days of the mission. The mind is, indeed, a wondrous thing!

CHAPTER 8

SLEEPING PILLS AND OVER-THE-COUNTER REMEDIES

LIMITING YOUR USE OF SLEEPING PILLS

I f possible, you should avoid sleeping pills and over-the-counter sleep remedies. They make you drowsy and appear to induce better sleep, but in reality they lead to disturbed, fragmented sleep. Natural sleep is always best and can usually be achieved by following good sleep strategies as discussed in previous chapters.

Used on a temporary basis, sleeping pills can be useful for known causes of short-term (transient) insomnia, such as jet lag, adjusting to shift-work rotations, the loss of a loved one, or the anticipation of a stressful event (some people say speech-

> Most sleep experts agree that sleeping pills should be used only in the smallest effective dose, for the shortest clinically necessary period of time, in specific situations, and with extreme caution. Many sleep specialists advise "never taking pills for sleep, period."[1]

making is second only to death as a feared event). Sleeping pills should never be used for chronic insomnia—lasting six months or more.

Many of the substances used in sleeping pills can be abused unintentionally. It is important to follow precisely the directions given by your doctor or pharmacist in the case of prescription medicine, or provided on the package of over-the-counter medications. Special attention should be paid to the relationship between the ingredients in the sleeping pill and any other medications that you may be taking. Certain medications could interfere with the effect of the sleeping pill and vice versa.

SOME DEFINITE DON'TS
Don't Ignore Label Instructions and Warnings.

Never take more than the recommended dose, never combine more than one prescription or over-the-counter remedy (unless told to do so by your doctor), and never drink alcohol while on sleep-enhancing medication. Be particularly careful if you are taking any other mood-altering drugs. Always inform your doctor if you are taking any other medications, prescribed by him or her or any other physician.

Don't Take Sleeping Pills If You Are Pregnant.

If you are pregnant, it is not a good idea to take sleeping pills. No sleep-enhancing medications have been proved safe for unborn infants. Also, pregnant women who are addicted to sleeping pills often give birth to infants who are also addicted. The newborn might experience withdrawal symptoms such as insomnia, irritability, hyperactivity, and tremors.[2]

Don't Drive.

Sleeping pills generally impair your ability to function in tasks that require you to be fully alert, so don't drive a car or do anything else that involves a high level of concentration and could be potentially dangerous.

Don't Take Sleep-Enhancing Medication Without Seeing Your Doctor.

Another important consideration if you are using or thinking about using pills to help you sleep is whether doing so may mask a medical or psychiatric cause to your insomnia. Many serious problems can express themselves through difficulty in falling or remaining asleep. So if you don't know why you're having trouble sleeping, an examination by your doctor before you begin taking sleeping pills might be a good idea. Otherwise you may never find the cause of your insomnia and may become dependent on medication to help you sleep. Worse yet, sleeping pills are respiratory depressants and can exacerbate sleep apnea and related illnesses.

TYPES OF SLEEPING PILLS

Sleeping pills are not all the same. They vary in potency, in the speed with which they are absorbed by the body, in the length of time they stay active, and in their side effects. These differences are important because they determine when you should take the pills and what to expect in terms of how alert you will be the next day or what other side effects you might experience. Your physician will be able to tell you which pill is the right one for you.

For example, if you need to be alert the next day, you want a drug with a short half-life, the amount of time it takes for half of the dosage to be completely metabolized by your body. A drug with a longer half-life might give you deeper sleep, but will have longer-lasting effects and will keep you groggy in the morning. Similarly, you don't want to take a rapidly absorbed drug two hours before going to bed or a drug that takes a long time to get absorbed at bedtime. And you may have to make a choice. Many of the less potent prescription drugs are safer than the stronger ones, so you may have to deal with a few extra side effects if the weaker drugs don't work for you.

The Benzodiazepines

The class of drugs that is most often prescribed for insomnia is the benzodiazepines. Benzodiazepines are relatively safe and have few side effects, but after long-term use you can become tolerant to them (meaning you'll need higher doses for the same effect) or dependent on them (meaning you won't be able to fall asleep without the pills).

Benzodiazepines are marketed as both tranquilizers (such as Valium) and sleep aids (such as Dalmane). The dosage is more telling of the function than is the name of the drug. For example, Valium is often used as a sleeping pill in higher doses and as a daytime antianxiety drug in lower doses.[3] Benzodiazepines can be prescribed by your doctor and cover a wide range of half-lives and absorption rates to meet the needs of many people.

Over-the-Counter Medications

Nonprescription, over-the-counter sleeping pills, such as Nytol or Sominex, are primarily antihistamines, as are many nasal sprays.[4] They will make you tired and groggy, but may also cause side effects if you are sensitive to them. Over-the-counter drugs should be taken with the same caution as prescription drugs. Always read the label so you know what you are taking.

THE DOWNSIDE OF SLEEPING PILLS

There are two *major* difficulties with sleeping pills:

1. Prolonged use brings tolerance; over time it will take greater and greater dosages of the same medication to have any effect on your sleep. Sleeping pills can be addictive and withdrawal will be difficult and uncomfortable.
2. After only two or three nights of using sleeping pills, when you stop you are likely to have bad dreams and even more insomnia than before.

In sum, for short-term insomnia, the safe and responsible use of both prescribed and over-the-counter sleeping pills can be an effective means of inducing sleep. Far better is to follow sleep strategies that will produce restful natural sleep night after night.

A DIGRESSION: WHAT'S THE MELATONIN CRAZE ALL ABOUT?

Relatively recently, synthetically produced versions of natural chemicals known to play a role in sleep have been marketed as sleep aids. Of these, the hormone melatonin has received the most attention.

Melatonin is naturally secreted by the pineal gland in the brain in response to darkness. It is known to lower body core temperature and cause drowsiness. Research has shown that melatonin taken as a supplement can hasten sleep onset and can reset your internal clock.[5] It does not seem to be addictive or produce the negative side effects (e.g., daytime grogginess) of many prescription sleeping pills and over-the-counter remedies.

A recent study showed that melatonin was particularly helpful in reducing insomnia in the elderly. We produce less melatonin naturally as we get older, so some researchers have suggested introducing melatonin using special slow-release capsules that mimic natural melatonin secretion.[6] The pill has also been touted as an aid in reducing jet lag if the right dose is taken at the right time.

Preliminary research indicates that low doses of melatonin (0.1 milligram) are as effective as high doses (10 milligrams) so more is not necessarily better. In fact, 1 milligram, the smallest dosage sold, is at least three times higher than the melatonin level normally found the human adult body.[7]

Perhaps melatonin, a potent hormone, can help "reset the body's aging clock, boost the immune system, keep cells from disintegrating, and slow the growth of tumors."[8] Perhaps not. There have been too few studies to date. Melatonin is not considered a drug and is therefore not screened by the Food and Drug Administration. Long-term effects have not been sufficiently studied and there is little regulation to check the production purity of melatonin or other remedies that you may buy in a health-food store. Some physicians and researchers are optimistic about melatonin supplements, but if you decide to take melatonin you are putting yourself at an unknown potential risk until long-term studies are completed.[9]

PART IV

COPING WITH SLEEP DEPRIVATION

You must sleep sometime between lunch and dinner, and no halfway measures. Take off your clothes and get into bed. That's what I always do. Don't think you will be doing less work because you sleep during the day. That's a foolish notion held by people who have no imagination. You will be able to accomplish more. You get two days in one— well, at least one and a half, I'm sure. When the war started, I had to sleep during the day because that was the only way I could cope with my responsibilities.

—WINSTON CHURCHILL

SLEEP

CHAPTER 9

THE NOD TO MIDDAY NAPS

IS IT A GOOD IDEA TO NAP?

t depends. If you have difficulty falling asleep at night you should avoid taking daytime naps. Your daytime sleep, especially if it's longer than thirty minutes, could be causing or exacerbating your nighttime insomnia. Most sleep specialists believe that naps are a good idea only if you can't manage to get one continuous period of sleep at night that is long enough to enable you to be fully alert all day long. If your hectic lifestyle doesn't permit you to get adequate nocturnal rest, take a nap on a regular basis. You won't be alone.

Half the world's population naps during the stretch from 1 to 4 P.M. in the afternoon.[1] Today, the average American takes about one to two naps a week. About 25 percent of Americans never nap, and about 30 percent nap more than four times a week. In a study of undergraduate students we found that 83 percent experienced mid-afternoon drowsiness and 81 percent take a nap at least once a week. Naps are healthy if you're sleep-deprived, as long as they don't interfere with your ability to fall asleep at your appropriate bedtime.

IS THERE AN INBORN TENDENCY TO NAP?

Recent research indicates that the human body is inclined to rest in the middle of the afternoon as well as at night, even after adequate nocturnal sleep. A heavy meal at lunch doesn't make you sleepy, it simply unmasks the physiological sleepiness that's

---

FAMOUS NAPPERS

- **Napoleon Bonaparte** napped because he was a chronic insomniac and could only sleep about three hours a night. (Notice the first three letters in Napoleon's name . . .)
- **Thomas Edison** napped in lieu of sleeping at night. He believed that sleeping was a waste of time, "a deplorable regression to the primitive state of caveman." But he napped frequently and for long periods.
- **Albert Einstein** felt that his daily naps "refreshed the mind" and made him more creative.
- During World War II, **Winston Churchill** scheduled his cabinet meetings around his daily catnaps.
- **Presidents Kennedy, Reagan,** and **Clinton** have been known to nap frequently.
- **Salvador Dalí** napped in his armchair, holding a spoon over a metal pan on the floor below. When Dali hit REM sleep and lost muscle tonus, the spoon would fall from his grip, bang the metal pan, and awaken him.

---

already in your body. The "postlunch dip" in alertness occurs whether or not food is consumed.

Dr. William Dement, director of the Sleep Disorders Clinic and Research Center at Stanford University, says, "It seems nature definitely intended that adults should nap in the middle of the day, perhaps to get out of the midday sun."

Dr. Dement, the Canadian sleep specialist Dr. Roger Broughton, and the University of Pennsylvania's Dr. David Dinges found that our natural sleep pattern is biphasic: We have a significant drop in body core temperature and alertness at night, and a similar but smaller drop in the middle of the day. According to Dr. Dinges, "A tendency to get sleepy at certain times [during the day] is biological." The afternoon trough occurs approximately twelve hours after the middle of your night's rest. It's then that you need a nap the most, especially if you have slept poorly the night before, and piled up a sleep debt.[2] It's not surprising that more accidents occur between 2 and 4 P.M. than in any other daytime hours. Better you should be napping than driving around in a zombielike state.

Some More Reasons to Nap

- Reduce stress! Siesta-loving Europeans and Latin Americans are more relaxed. They usually score better on stress tests than North Americans.[3]
- The risk of heart disease is shown to be greatly reduced by regular thirty-minute naps.[4]
- Naps greatly strengthen the ability to pay close attention to details and to make critical decisions.
- Naps taken about eight hours after you wake have been proved to do much more for you than if you added those twenty minutes onto already adequate nocturnal sleep.

HOW LONG SHOULD A NAP LAST?

A nap should be about fifteen to thirty minutes in duration. If you nap longer than thirty minutes your body lapses into delta, or deep, sleep. Delta sleep is difficult to wake from and if interrupted or just completed, can leave you feeling terribly groggy. If you are severely sleep-deprived and must nap longer than thirty minutes you should probably extend it to a full hour and a half to complete a sleep cycle. But don't expect to be fully alert until at least an hour after the nap. Your nocturnal sleep time is likely to be shorter after a long daytime nap. Count the time spent napping as part of your total sleep quotient.

Dr. Jeffrey Midgow contends that the body only needs about fifteen minutes to nap because it "is a very resilient system that doesn't need much more than that to rejuvenate. If you take time to turn the nervous system off the whole system recharges."[5] Your mood will improve, as will your alertness.

If you are going to nap in the middle of the day, be consistent and make a habit of napping every day. An irregular napping schedule might disrupt your internal body clock and nocturnal sleep pattern. Napping only on the weekends is like dieting or exercising only on the weekends to make up for a week of overeating and not exercising—it doesn't work. Brief naps taken daily are far healthier than sleeping in or taking very long naps on the weekend. If you nap too long on Sunday afternoon to catch up from your week's hectic schedule, you'll have trouble getting to sleep that night. This shifts your biological clock, making it difficult to get up in time on Monday to start the new workweek.

You might have to work through your mid-afternoon period of drowsiness because there is no time or opportunity for a nap. After this you might feel increased alertness, especially if you're doing something interesting. This is because your clock-alerting circadian rhythm has passed its low point and is beginning to rise again. However, don't think this second wind means you're not still prone to an unexpected sleep seizure. You haven't repaid any previous sleep debt. If you start doing something sedentary, like driving a car or watching television, you could suddenly fall asleep.

Late-afternoon napping isn't healthy. It delays your falling-asleep time in the evening and begins to shift your biological clock, making getting up in the morning a struggle. Senior citizens who have nocturnal insomnia should refrain from napping at all. It will only make things worse. Better to be exhausted in the evening and get a long, continuous nocturnal sleep.

You might want to take a mid-afternoon nap but can't seem to fall asleep. If that's the case, don't worry. Maybe you're not overly sleep-deprived in the first place; or you might be too stressed or stimulated to nod off. Just the act of relaxing, meditating, or lying down with your eyes closed will be beneficial in restoring energy.

> **IS NAPPING NATURAL?**
> The judicial system of the United States has made it official: Napping is natural. A recent ruling by the U.S. Court of Appeals for the Second Circuit granted a new trial to a defendant who claimed his attorney was asleep for most of the proceedings. The court declared, "There are states of drowsiness that come over everyone from time to time during a working day."[6]

PROPHYLACTIC NAPPING

Napping makes it possible to deposit needed assets in your sleep bank account. If you know you'll be going to bed late on a given night because of travel or evening social events, take a preparatory nap that day. These prophylactic naps can be two to three hours in duration and are effective in providing addi-

ADVICE FOR PARENTS OF NEWBORNS
Exhausted parents of newborns should take naps when their infants nap. Don't try to use the free time to catch up on household chores or work if you're already seriously sleep-deprived. A well-rested caregiver is a better caregiver.

tional hours of alertness that night. You'll feel a bit groggy for about thirty minutes following a long nap. That's called "sleep inertia," but it will soon give way to feelings of alertness.

Researchers have demonstrated that when people napped in preparation for an all-nighter, their performance the next day was superior to that of those who did not take the prophylactic nap. In fact, they were 30 percent more alert and optimistic than people who did not nap. When you take a long nap, remember that your nocturnal sleep will be reduced by the amount of shut-eye that you got that day.

Napping for fifteen minutes every four hours, for a total of one and a half hours sleep per day, first suggested by Leonardo da Vinci, is often used by solo sailors on long voyages and for those performing long hours under emergency circumstances (fire and earthquake rescue, for example).[7] As a stopgap measure it seems to work for short periods of time.

CAN YOU GET THE NOD AT WORK?

The notion of sleeping on the job isn't very well received by industry. Companies equate napping with laziness, and lazy or not, you might get chastised or even fired if you're caught trying to catch a few winks. But the fact is that today nearly everyone is sleep-deprived, and sleepy workers are irritable, likely to make mistakes and cause accidents, and more susceptible to heart attacks and gastrointestinal disorders. That costs money and disrupts lives.

For hundreds of years in Latin America and in Europe, everyone quit work for a couple of hours in the mid-afternoon and went home for a refreshing nap. However, siestas are becoming a thing of the past. As countries become more industrialized and transportation to and from work more congested, the siesta has all but disappeared. Too bad. Time for a change!

A coffee or cola break, perceived as a legitimate part of the workday, does provide momentary feelings of alertness. But consumption of caffeine will be followed by feelings of lethargy and reduced REM sleep that night. A debt in your sleep bank account, the cause of midday sleepiness, is not reduced by these artificial stimulants. Why not attack the problem directly and get some needed sleep? We must strive to recognize brief naps as legitimate, and much more valuable than coffee breaks.

The corporate culture will gradually change as information on the tremendous costs of sleep deprivation, and on the biphasic pattern of natural sleep, become more widespread. Accepting the concept of napping will reduce errors and accidents, raise job satisfaction, reduce illness, and ultimately improve the bottom line. What companies lose in time they can make up in increased productivity.

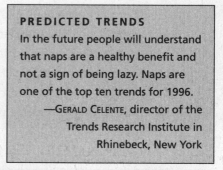

PREDICTED TRENDS
In the future people will understand that naps are a healthy benefit and not a sign of being lazy. Naps are one of the top ten trends for 1996.
—GERALD CELENTE, director of the Trends Research Institute in Rhinebeck, New York

It's a win-win situation that should make sense to everyone.

Since corporations have accepted the concept of "power breakfasts," I coined the term "power nap" to encourage institutionalization of naps at work. A recent survey indicated that many executives take brief naps in the office to recharge their batteries. Why not allow all workers to have the same privilege? In many corporate offices comfortable couches and sleep-friendly office furniture for midday snoozes are replacing hard-backed chairs. Cots are becoming available in lounges off the factory floor. And it's working wonders. Employees are more alert, more productive, and less accident-prone.

SOME HELPFUL HINTS FOR NAPPING AT THE OFFICE

If you've convinced the boss, or if you are the boss, take a nap on your work break. Here's how to make naps work:

- Get rid of all the distractions! Turn off the ringer on the phone, close the door, and turn off the lights. Put a sign on your door knob: "I'm recharging my batteries," or "I'm power-napping."
- Avoid caffeine after that first morning cup of coffee. Otherwise you may feel jittery and unable to fully relax.
- Consider your environment. Furniture has recently been designed to aptly suit office naps, including power-nap executive chairs that recline, massage your head, and envelop it in darkness. There are desks that transform into cots, and chairs and couches that have built-in audio decks for playing relaxing music and alarm clocks to wake you up.

    You don't have to go to extremes. If you've got a couch, use it! Lying down is the optimal position. If not, lying back in a chair with your feet up is the next best way. Have a pillow handy so you'll have a comfortable headrest. Or you can just sit at your desk and put your head down for a few minutes' rest.
- Schedule regular rest periods. Just as with nocturnal sleep, it is important to get your body used to napping at about the same

time every day, for about the same length of time. Most people tend to feel sleepy about eight hours after awakening.

- Even on days when you don't feel particularly sleepy, try to take a rest around nap time instead of taking a coffee break.

- Limit your nap time. A fifteen- to thirty-minute nap is ideal. Many tired executives worry that if they put their heads down they will fall asleep and not wake up for hours.

Figure 9.1. Westclox Napmate.

Not to worry. "Power-nap" alarm clocks are available with a one-button preset for naps of specific duration (see Figure 9.1).

- If your work includes air travel, which can easily contribute to stress and exhaustion, particularly if you cross time zones, consider napping during your flight. The twenty-minute periods you spend not working will pay off in hours of more efficient and productive thinking. A short nap before arrival at your destination is also good, especially if you're planning to drive a rental car.[8]

"He's back from his power lunch, now he's taking his power nap."

From *The Wall Street Journal*. Permission, Cartoon Features Syndicate.

Forty million Americans are now working full or part time in the home and a good portion are probably taking naps. The emphasis is going to be on what is produced, not on where you work or at what time of the day or how long you work. Napping should not be frowned upon at the office or make you feel guilty at home. It should have the status of daily exercise.

PROACTIVE EDUCATION

Showing how naps and good sleep strategies can quickly restore alertness, enhance performance, reduce mistakes and accidents, and affect profits is a quick way to grab any CEO's attention. Talk to your boss, give him or her some literature or tapes on sleep deprivation and performance, or invite a sleep expert to give a seminar at your company.

WHEN YOU'RE SLEEP-DEPRIVED, DO YOUR MIND AND BODY A FAVOR. TAKE A NAP.

"Damn it, Abernathy, it's nap time."

Drawing by Weber; © 1997 The New Yorker Magazine, Inc.

CHAPTER 10

SURVIVING AS A SHIFT WORKER

S hift work became commonplace in America following
World War II. Increasing competition and rising costs made
it difficult for many industries to justify operating factories
only forty hours per week—less than one fourth of capacity. A 168-hour round-the-clock work week was instituted that
allowed for continuous operation and quadrupled output.

Hospitals, police and fire departments, and the post office
have long operated on a twenty-four-hour basis; now factories,
grocery stores, and many other businesses also never close.
Some must work while others sleep. Today 22 million people,
one fourth of America's workforce, are shift workers.[1]

Some people work only at night, and others rotate between day, evening, and night shifts. Although there are more than 700 variations of rotating shift schedules, the most common is the "seven swing." With this schedule there are seven straight eight-hour day shifts, then two days off; seven eight-hour night shifts, then two days off; followed by seven eight-hour evening shifts, then two days off. "Such a life," Dr. Martin Moore-Ede of Harvard Medical School maintains, "represents the equivalent of continually spending a week on Boston time, followed by a week in Paris, followed by a week on Tokyo time."[2]

Needless to say, this system almost guarantees that shift workers will be in a state of perpetual jet lag, without ever having had the pleasure of leaving home.

HOW SHIFT WORK AFFECTS PRODUCTIVITY

Traditional shift schedules assume that people can adapt to most work cycles. In reality, employees are pushed far beyond what the human body can reasonably withstand.[3] Shift workers work 400 more hours a year than those who work only 40 daytime hours (the equivalent of ten additional 40-hour weeks).[4]

> More than half of all shift workers admit to falling asleep on the job at least once a week.[5]

Trying to work and remain alert at a time when the brain is calling for sleep is futile. Our natural sleep-wake cycle, regulated by light and darkness and programmed over thousands of years of evolution, prevents us from easily adjusting to night or rotating shifts and irregular work schedules. If we operated machinery or equipment in the same manner as we "operate" our shift workers, we would be accused of reckless endangerment.

Proneness to Errors

Sleepy shift workers are prone to making errors that result in poor-quality work and reduced productivity. They are *forty* times more likely than day workers to be involved in accidents—at work, on the highway, and at home.[6]

Police fall asleep while stopped at traffic lights and report being awakened by the horns of cars waiting behind them. In a recent survey of police officers, 80 percent reported they had fallen asleep once a week while working the night shift.[7]

Sleep-deprived assembly-line workers fall off their stools, batches of defective products slide past dozing inspectors, and exhausted forklift operators crash their machines into walls.[8]

Exhausted resident physicians working the night shift report falling asleep while taking patient histories. Some have hallucinated while doing surgical procedures. Nearly half report falling asleep at the wheel on the way home.[9]

The Financial Toll

The financial toll in terms of productivity and safety that results from reduced alertness is more than $70 billion per year.[10] For example, the launch of the shuttle *Columbia* on January 6, 1986, almost resulted in a tragedy because of operator fatigue. Technicians had been working twelve-hour night shifts for three consecutive days. In a sleep-deprived state, one operator inadvertently drained 4,000 pounds of liquid oxygen from the shuttle external tank just five minutes prior to the scheduled launch. Luckily, the mission was aborted just thirty-one seconds before liftoff, because of a secondary effect on the engine-inlet temperature. The liquid oxygen loss was undetected until after the postponement.

The crew of the *Challenger* shuttle were not as fortunate.

Sleep researchers attribute questionable last-minute evaluations of the reliability of O-ring seals to the insufficient sleep and irregular hours of NASA managers involved in the decision to launch. Two of the three top managers had had less than three hours of sleep for the three consecutive nights prior to the catastrophic mission.

HOW SHIFT WORK AFFECTS TIME OFF

When shift workers manage to get a weekend off, the accumulated fatigue caused by constantly changing work hours leaves them exhausted and in a state of near collapse.[11] It's difficult for them to get needed sleep and fully participate in family and social activities.

- Shift workers find it difficult to coordinate their free time with their children, who are either in bed or at school.
- Shift workers report problems in fulfilling their sexual roles with their spouses, often leading to marital dissatisfaction.
- Shift workers must leave evening social events early in order to get to work on time.[12]
- Even with as much as 30 to 40 percent more pay than their daytime counterparts, shift workers consistently rate their job satisfaction lower than do day workers, and they keep looking into getting a day job.[13]

HOW SHIFT WORK AFFECTS SLEEP AND HEALTH

According to the American Sleep Disorders Association, 5 million people endure sleep disorders as a result of shift work.[14] Researchers at the Institute for Circadian Physiology in Boston, Massachusetts, have found that **70 percent of shift workers** have trouble falling asleep. Shift workers average one to two hours less sleep than their daytime counterparts during the week, and three or four hours less on weekends.[15]

Alcohol, Tobacco, and Caffeine Abuse

Shift workers report an increased use of alcohol to induce sleep (a dubious technique), and excessive smoking and caffeine intake to aid in alertness. One twenty-year-old hospital orderly developed sleep paralysis after working the midnight to 8 A.M. shift five nights a week for eight months. He would awaken during the night and be unable to move for up to fifteen minutes. He had excessive fatigue during the day, which he tried to combat by taking three and a half hours of naps per day and drinking four cups of coffee and two cola drinks daily.[16]

General Physical and Mental Well-being

- Shift workers report more sick days, have more heart attacks and cardiovascular and gastrointestinal diseases, and suffer more mood disorders, depression, and psychiatric problems than their counterparts on regular day shifts.[17] Some studies have shown that shift work may actually decrease one's life span.[18]
- Shift workers' mental well-being may also be affected by their work schedules as evidenced by tension, nervousness, and irritability.[19] This becomes more pronounced with age. Also, it is more difficult for older people to maintain sleep during the day than it is for younger people.[20]
- Shift workers who miss three or more hours of sleep decrease their resistance to viral infection by as much as 50 percent. Dr. Michael Irwin, a psychiatrist at San Diego Veterans' Hospital, found that people who were forced to stay up later than usual showed a decrease in the activity of natural killer cells, cells in the blood that fight viral disease. When rest is interrupted, so are the

essential immune-building secretions usually emitted during sleep.[21]

- Sleep experts recommend that people with a history of digestive tract disorders, diabetes, and epilepsy should avoid shift work.[22]

HOW EMPLOYERS CAN HELP EMPLOYEES STAY ALERT

Employers can maximize productivity as well as reduce shift workers' sleepiness and increase their job satisfaction by taking the following steps:[23]

- Provide their employees education about biological clocks, circadian rhythms, and the importance of good sleep strategies.
- Change shift schedules so workers (1) rotate no more frequently than once every three weeks, with two days off in between shift changes; and (2) rotate in a clockwise direction, from the day shift to the evening shift to the night shift.
- Provide adequate breaks and avoid excessively long shifts.
- Provide areas for napping.
- Provide bright daylight-spectrum lighting in work areas.
- Keep to a minimum the number of workers on night shifts.[24]

STRATEGIES SHIFT WORKERS CAN USE TO IMPROVE THEIR SLEEP

Whether or not your company has a shift schedule that is "friendly" for your biological clock, here are some tips that you can follow to make the most of your days and nights:

- Take a nap two hours prior to your shift to help make up for sleep loss.[25]
- Try to prepare for your sleep schedule on your days off before the shift changes. For example, if your next shift is an evening shift,

stay up a little bit later and then sleep later in the morning.[26]

- Stay physically fit. This reduces overall fatigue and also increases alertness on work shifts.[27]

- During the evening and night work shifts, expose yourself to full daylight-spectrum bright light. Bright lights (at least 2,500 lux) that mimic the sun's spectrum and intensity can help reset the body's sleep and wake cycles.[28] These lights are available in hardware stores and nurseries under brand names such as Vita-Lite and Growlux.

- Eat a meal or snack at the same time each day for the duration of the work shift to establish some regularity for your body's internal clock.[29]

- Avoid caffeine during the last half hour of your shift because it can take so long to leave the body that your ability to fall asleep and remain asleep can be affected.[30]

- On your morning drive home after working all night, try wearing dark glasses so the daylight will not reset your biological clock and delay your sleep cycle.

- Avoid alcohol after work if you must sleep during the day; any alcoholic drinks that you consume to help you fall asleep can actually cause a highly disruptive sleep pattern during your sleep period. The resultant drowsiness may also lead you to take amphetamines or caffeine to stay awake at work, and a vicious cycle begins.[31]

- Try to relax, unwind, and go through a regular bedtime routine before sleeping. Some people choose to relax a little, eat a light snack, and then go to bed.[32]

- Pay special attention to your sleeping environment. Ensure adequate screening from noise and interruptions by keeping your room quiet and by insisting that family and friends respect your sleep hours. Remind friends that a phone call or visit during mid-afternoon is about as welcome as a middle-of-the-night knock on the door is for them!

Figure 10.1. The Sunlight Box, available from the SunBox® Company, 19217 Orbit Drive, Gaithersburg, MD 20879.

- Keep your bedroom cool and pitch-dark. It may be necessary to invest in extra-thick drapes or some other means of blocking out the bright daylight.[33]
- You might consider buying a sunlight box to help you adjust to rotating schedules. Exposure to daylight-spectrum light before going on the evening or night shift will help reset your circadian clock (see Figure 10.1).
- Try to schedule doctor and dentist appointments that work for *your* schedule. For you, an appointment at noon may be like an appointment at 3 A.M. for a day worker.[34]
- Try to maintain a regular sleep schedule for the duration of the

week. Get up at the same time seven days a week and also keep a regular meal schedule when possible.[35]

• If problems persist, seek a physician. See Appendix D for a list of approved sleep clinics in your area.

A DIGRESSION: EFFECTIVE SHIFT-WORK SCHEDULES

• Schedules that take human physiology and biological clocks into account result in major improvements in workers' alertness, sense of well-being, and job performance. Dr. Martin Moore-Ede of the Institute for Circadian Physiology in Boston, Massachusetts, claims as much as 30 percent improvement in productivity with well-designed schedules.[36]

• Dr. Charles Czeisler of Harvard Medical School devised a new schedule that rotated shifts clockwise (day to evening to night) and slowed rotations down from every week to every three weeks for the workers at Great Salt Lake Minerals & Chemicals. More than 70 percent of the workers preferred the new schedule, there was a huge drop in the number of complaints about sleep and stomach troubles, and there was a 20 to 30 percent increase in work production.[37]

• Dr. Czeisler also helped the Philadelphia Police Department take circadian rhythms into account in designing their schedules. Major revisions included cutting shift changes by one third, reversing the direction of rotating shifts from counterclockwise to clockwise, and decreasing the number of consecutive days worked. Officers reported a fourfold increase in the frequency of better sleep, a 25 percent decline in the number of sleeping-on-duty episodes on the night shift, and a 29 percent increase in alertness. Officers had 40 percent fewer on-duty automobile accidents per mile driven than in the preceding two years, and they reported using fewer sleeping pills and less alcohol to alleviate the consequences of sleep deprivation.[38]

- A better shift schedule not only improves worker productivity and morale, but also offers substantial financial savings. For example, an oil refinery saved $2,437,000 per year by changing from an outdated traditional schedule to a more efficient schedule:

  $900,000 saved in reduced overtime payments
  $600,000 saved in reduced operator idle time
  $500,000 saved in maintenance idle time
  $270,000 saved in improved utilization of seven-hour shift
  $86,000 saved in reduced absenteeism
  $81,000 saved in improved health and safety[39]

CHAPTER 11

REDUCING TRAVEL FATIGUE

OVERCOMING JET LAG

Whether for business or pleasure, many of us increasingly rely on air travel to reach our destinations. But flying can be exhausting, and flying across multiple time zones can cause jet lag.

Jet lag is more than just being tired from traveling. Technically called *circadian dysrhythmia*, jet lag is a disruption of the body's intricate biological inner-sleep cycle caused by crossing multiple time zones quickly. It occurs when the clock on the wall at your destination indicates a time far different from your internal body clock, which is still back home on its original time schedule. You are out of sync with your environment, and that has ramifications for your alertness and feelings of well-being. Jet lag causes business travelers to be less productive, athletes to be less sharp, and tourists to be too tired to fully enjoy their long-anticipated faraway vacations.

Jet lag is not new.

John Alcock and Arthur Brown, pilot and navigator of the first nonstop transatlantic flight in 1919, experienced jet lag. Alcock reported, "In our eastward flight of 2,000 miles we had overtaken time, in less than the period between one sunset and another, to the extent of three and a half hours. Our physical systems having accustomed themselves to habits regulated by the clock of Newfoundland, we were reluctant to rise at 7 A.M.; for our subconsciousness suggested that it was but 3:30 A.M."[1]

Greg Louganis, a gold medalist on the U.S. Olympic Diving Team, reported that the reason he struck his head on the ten-meter platform during a reverse dive at the Olympic trials in Moscow in 1979 was that he had been severely affected by jet lag.[2]

West Coast National Football League teams beat East Coast teams more often and by more points in *Monday Night Football* games. Whether at home or away, West Coast teams are playing closer to their peak athletic performace time (late afternoon) with start times of 9:00 P.M. Eastern Standard Time.[3]

The former secretary of state John Foster Dulles felt that his decision on the controversial Aswan Dam in Egypt was one of the great mistakes of his career, and that he might have taken a more conciliatory stance with the Egyptians had he not been so weary from jet travel.[4]

Lowell Thomas was a renowned radio and film commentator and world traveler. He was first to broadcast from an airplane and from a ship, and his sonorous baritone reached audiences of billions during his distinguished forty-five-year career. In 1963, Thomas was hospitalized for muscular tremors, extreme fatigue, fainting, and vertigo. Initially he was thought to have had a heart attack, but Thomas himself realized that since he had been continually crisscrossing the globe in pursuit of news stories, crossing all twenty-four time zones at least twice in recent months, his maladies were actually caused by a severe case of jet lag.[5]

SYMPTOMS OF JET LAG

Daytime sleepiness. Ninety percent of travelers report experiencing daytime fatigue and sleepiness. If you give in to the urge to sleep during the day at your destination, you may not be tired enough to sleep at bedtime.

Insomnia. The next most common symptom of jet lag is insomnia. You experience difficulty falling asleep at night. Once you do get to sleep, you'll have less deep sleep and less REM sleep. The night's sleep is often fragmented by frequent awakenings.

Poor concentration. More than two thirds of air travelers report having poor concentration, or in severe cases of jet lag, temporary amnesia. You become unable to focus attention, cannot think clearly, and have foggy memory, and your ability to write coherently is impaired.

Disorientation. Many travelers also experience disorientation. You become confused and cannot remember where you are, especially when you wake up in the middle of the night.

Slower reaction time. Many travelers suffer from slower reflexes. This seems especially relevant if you must cope with unfamiliar traffic patterns (like driving on the opposite side of the road) in your new destination.

Gastrointestinal problems. About 50 percent of travelers say that jet lag disrupts their digestion. You might have a poor appetite or have hunger pangs at odd hours. You might become constipated and experience heartburn or ulcers from eating meals at hours when you would usually be asleep.

Other symptoms. Other reported symptoms of jet lag include irritability and depression, headaches, urinary system disruptions, alterations in the menstrual cycle, tendency to catch colds, and changes in the effectiveness of medicines.[6]

WHAT FACTORS INFLUENCE YOUR SUSCEPTIBILITY TO JET LAG?

Number of time zones crossed. Jet lag starts to be noticeable when you cross more than three time zones. The greater the number of zones crossed, the greater the severity of jet lag symptoms.

Direction of flight. The direction of your flight matters greatly. When you fly eastbound, or against the direction of the sun, jet lag tends to be more severe than when you fly west. You can actually take 50 percent longer to recover from jet lag after an eastward flight than after a westward flight of the same distance. When flying westward, you are allowing your body to follow its natural inclination to extend the day; remember, the body clock's natural sleep-wake cycle is around twenty-five hours, not twenty-four. Because north-south and south-north flights do not involve time zone changes, they do not cause jet lag. You might feel physical or mental exhaustion after a long flight in these directions, but you will not be jet-lagged.[7]

Age. The older you get, the more you are likely to experience the debilitating effects of jet lag.[8] Babies under age three seem unaffected, children adapt better than their parents, and the elderly seem to have the most trouble.[9]

Personality characteristics. Sensitivity to jet lag appears to be linked to certain personality characteristics:

- If you are extroverted, easygoing, and enjoy the company of others for travel or meals, etc., you are likely to suffer less from jet lag.[10]
- If you are calm in most difficult situations, it may be easier for you to adapt to changes in your body clock.
- Night owls typically fare better than morning larks when flying west, but the early-rising larks seem to cope better when flying in an easterly direction.[11]
- If you exercise regularly and are in good health, jet lag's effects are likely to be lessened.
- If you are very regimented in your living habits (rising, eating, going to bed at the same hours each day), you may suffer less from jet lag than if your schedule is more irregular.[12]
- If you are determined to fight jet lag, it may not affect you as much. Margaret Thatcher is a good example of a person who does not in general suffer from jet lag; she has convinced herself that it won't get her down.[13]

Sleep debt. The amount of sleep debt you are carrying can affect your susceptibility to jet lag. In general, the better rested you are, the better you'll fare when faced with jet lag.

HOW TO COMBAT JET LAG

In an Upjohn Company survey of experienced travelers, 94 percent reported they were bothered by jet lag. However, only about half reported taking any measures to avoid its debilitating symptoms.[14] The following tips, compiled by a host of travelers and researchers, will help ease the stress and fatigue from any lengthy air travel, and should moderate the biological clock-resetting process necessary to counter jet lag.[15]

1. Planning Your Flight

- Avoid selecting a flight with a morning departure so early that you have to lose sleep in order to get to the airport on time.

- If feasible, plan to arrive at your destination in time for a full night's sleep.

- Avoid red-eye flights. Although you may save some time and the cost of another night in a hotel, the cost of losing sleep can be even greater, in terms of mood, health, and performance.

- In preselecting a seat, request one away from bathrooms, the galley, or where young infants are often placed.

- Be aware of which side the sun will be on, and try to sit on the opposite side.

- Consider a seat location that will afford you the greatest leg room, such as the emergency row or on the aisle. Make sure you are seated in a row with adjustable seatbacks.

- Give yourself time. Plan ahead and arrive at the airport with plenty of time to spare and a chance to upgrade to roomier and quieter seating areas.

- At the gate, ask the boarding agent if there are any completely empty rows. Three empty seats makes a quasi-bed.

- Pack a small bag of items to help you stay comfortable on the plane. Include an eyeshade, earplugs, slipper socks, gum (for equalizing ear pressure on takeoff and landing), moisturizer, lip balm, a nasal decongestant, and a bottle of water.

- Plan to wear loose-fitting clothing on the flight and dress in layers for warmth and comfort.

- Start to preset your biological clock five days before you leave: If flying east, start going to bed and waking up earlier each day; if heading west, stay up later and get up later.

- Stay calm while preparing for your trip, to ensure a relaxed departure. Don't leave trip preparations until the last minute. Be well rested, not exhausted, when you start your journey.

2. During Your Flight

- Ask for a pillow and blanket as soon as you board. Many flights do not carry enough of a supply to go around.

- As soon as you sit down on your flight, change your watch to the time at your destination and begin living by that time—acclimatizing yourself to that time zone.

- Drink lots of water and juices to counter dehydration from the dry cabin atmosphere. Dehydration can retard the process of resynchronizing your biological clock with destination time.

- Avoid alcohol. Cabin pressure at higher altitudes raises your blood-alcohol level such that two drinks at high altitude are as potent as three drinks on the ground. Alcohol tends to exacerbate the dehydration problem and interferes with your body's ability to process oxygen. Perhaps most significant of all, alcohol will disrupt REM sleep and fragment sleep throughout the night.

- Avoid smoking, overeating, or eating spicy foods, all of which may interfere with your sleep.

- Take several strolls down the aisle to improve blood circulation.

- Do some stretching. While in your seat try putting your feet on your hand luggage to lift your thighs away from the edge of the seat and allow freer circulation.[16] Try deep breathing to replace and refresh the air in your lungs and bring fresh oxygen to your blood. Stretch your arms above your head as if you were picking items off a high shelf, and rotate your head right and left, then up and down, to relieve the tension in your neck muscles.[17]

- Loosen your clothing as an aid to circulation. Take off your shoes.

- If you wear contact lenses, consider removing them while in flight so that your eyes do not become irritated because of the extremely dry atmosphere in the cabin.

- While airborne, eat and sleep according to your new schedule, not the airline's imposed schedule. Even though it's still daytime outside the plane, if it's nighttime at your destination, forget the movie and meals and get some sleep. Put on your eyeshades, put in your earplugs (or put on headphones), and tell the cabin crew you do not wish to be disturbed by meal service. Covering yourself with a blanket helps keep you comfortable as your body temperature drops from inactivity and sleep.

- If necessary, consider using an antihistamine to induce sleep. Melatonin (discussed at the end of Chapter 8, "Sleeping Pills and Over-the-Counter Remedies") may be promising for reducing the effects of jet lag. Because inappropriate dosages can produce mood-altering side effects, and because we don't know melatonin's long-term effects, you should check with your physician before using the drug.

- Use daylight to help reset your biological clock to destination time. You can simulate daylight during your flight, even at night, with visors that have daylight full-spectrum bulbs in the shield which shine obliquely on the eyes (see Fig 11.1). By reading with the visor in place for fifteen to thirty minutes at appropriate times (as suggested in the manuals for such devices), your biological clock will begin to make the proper adjustment toward your new time schedule, even before you land.[18]

Figure 11.1. Bio-Brite Light Visor, available from the SunBox® Company, 19217 Orbit Drive, Gaithersburg, MD 20879.

- If you will be flying long distances eastward at night, do wake up and have a good breakfast at 7 A.M. local time, then stay awake and walk around on the plane while you can.

3. Checking Into Your Hotel

Remember that the ideal bedroom environment for good sleep hygiene is quiet, dark, cool, comfortable, and secure. The better your hotel accommodation meets these criteria, the better your chances are for a good night's sleep. Some hotels have designed special rooms for weary travelers. Hilton has "Sleep Tight" rooms in some properties, complete with special soundproofing, sleep gadgets, and minibars stocked with sleep-inducing snacks. In any case, be a wise traveler:

- You can limit noise by reserving a room on a high floor if on the street side of the hotel. Request that your room not be close to elevators, stairways, vending or ice machines, and hospitality suites. If that's not possible, prepare to use your earplugs. Sometimes the air conditioner fan can be used to mask unwanted noise.
- Ask for a room with an eastern or southern exposure for more morning sun, making it easier to become alert in the morning. If south of the equator, get a room with an eastern or northern exposure.
- If you aren't pleased with the location of your room, ask for a room change before you unpack your bags.
- Pull the heavy drapes closed at night to keep out city light and reduce noise.
- Keep the room at 65° F. during the afternoon and night. Check the thermostat as soon as you arrive and call the management if there is a problem.
- Request extra pillows or blankets when you check in. Maid service is sometimes hard to come by late at night. If you have a special pillow that nearly always ensures sleep, bring it along.
- Pack a nightlight and plug it in so you can navigate the room without turning on a bright light.

- Before you turn in make sure the door is bolted and a "Do not disturb" sign is on the outside doorknob.

Set your alarm clock and leave a wake-up call request with the hotel operator. This provides double insurance that you'll wake up on time. Ask the operator to hold all calls until morning. Turn off the light, knowing you're prepared for a peaceful night's sleep.

4. Day One at Your Destination

- On arrival, follow the meal pattern and sleep-wake schedules appropriate at your destination. If you've flown eastward and it's still the middle of the night according to your biological clock, yet morning according to the time at your destination, don't go to bed at the hotel for a few hours, even though you're exhausted. It will only delay the necessary resetting of your internal clock. Hoteliers report that eastbound travelers who intend to "take a short nap" because they arrive early in the morning after an all-night flight often sleep for six to eight hours if not awakened by an alarm clock or a call from the front desk. So much for the first day at your new location. . . . It's far better to push yourself through that first day and fall into bed early that evening, exhausted but ready for a good night's sleep on local time.
- If you've flown eastward and arrive in the early morning, try to get outside in the sunlight as soon as possible. Daylight is a powerful stimulant for regulating the biological clock, and staying indoors actually worsens jet lag.[19]
- If you've flown westward and it's already evening according to your biological clock, yet still afternoon at your destination, spend time outdoors in the afternoon sun. It will help delay your biological clock, getting it in sync more quickly with local time. Remember, it's easier to adjust to a westbound time change than an

eastbound change because of the circadian rhythm's tendency toward a longer, twenty-five-hour day.

- Getting some exercise, even a brisk walk, after a long flight will raise your endorphin levels. This in turn will reduce stiffness and pain, relax your muscles, help suppress your appetite, and create feelings of optimism and happiness.

- Business executives, government officials, and athletic teams should delay doing business or engaging in sports until the second day abroad after more than a five-hour time shift. Otherwise, mistakes will be made, negotiations will suffer, and games will be lost.

All of the above suggestions should minimize the burdensome effects of adapting quickly to a distant time zone. If you're still miserable for several days, perhaps next time you should think about traveling by car, bus, train, or ship. When you cross multiple time zones slowly your biological clock can handle the gradual time changes quite easily. In the good old days there was no such thing as jet lag. Life was slower, travel was slower. There was less insomnia. People were more alert. Times have changed, but progress does not always make perfect.

ASLEEP AT THE WHEEL

Sleepiness on the road is a serious problem. If you're a busy person carrying a sleep debt, the odds for accidents and catastrophes are higher than you might think. Most of us drive every day, so familiarization with the following information could easily save your life and the lives of others sharing the highway, or at least prevent an accident. Let's look at some facts:

- Nearly one in every three Americans, 31 percent of the population, report having fallen asleep at the wheel. At least one in every twenty Americans has caused an accident by doing so.[20] Sleepiness is second only to drunkenness as a cause of automobile accidents, and sleep experts warn that drowsy driving may be just as prevalent and as dangerous as drunk driving.[21] According to the American Medical Association, sleepy drivers are "America's nightmare."

- The National Highway Safety Administration estimates that at least 100,000 crashes, 71,000 injuries, and 1,500 fatalities each year are caused by drivers who fall asleep at the wheel.[22] This represents 30 percent of all highway vehicle accidents.
- Sleepiness and fatigue is the number one cause of heavy trucking accidents.[23] One out of every five victims is in the cab of the truck; the remaining four victims are innocent pedestrians or motorists.[24]
- Approximately 30 million Americans suffer from sleep apnea, a disorder marked by repetitive pauses in breathing during sleep, and daytime sleepiness (see Chapter 13, "Insomnia and Beyond"). Ninety percent of those with the disorder have not been diagnosed and are untreated. A study by Terry Young at the University of Wisconsin found that men with the disorder get into three times as many automobile accidents as the rest of the population. Men and women with undiagnosed sleep apnea are seven times more likely to have multiple accidents.[25]

We may feel our eyelids getting heavy or have difficulty concentrating on the road, but we don't pull over. Instead, we try to "drive through" the sleepy episode, and even rush our trip by speeding up.[26] Dr. David Dinges, a sleep expert at the University of Pennsylvania, reports, "Even if you don't fall asleep, when you drive drowsy, you drive impaired. Your reaction time is slowed, your perception is distorted and you don't stay in your lane as easily."[27]

Stimulation from business meetings or socializing generally masks our sleepiness. It's when we get in the car to go home that the urge to sleep can be sudden and overpowering. Many drivers insist that they can tell when they are about to fall asleep, but research shows otherwise. Most drivers who fall asleep do so without realizing it until it's too late. Sleep expert Dr. William Dement says many of us experience "microsleeps," or "uncontrollable and unpredictable bouts of sleep that happen 'faster than a seizure.' "[28] We may fall asleep for only a few seconds, but behind the wheel of car, a lot of damage can be done in that instant.

For example, a South Dakota driver was thinking he should stop for coffee to keep awake. Moments later he slammed into a truck that had stopped at a crossroads; a steel girder from the truck bed smashed into his car and decapitated his wife. In July 1994, twenty-three people were burned in their homes in White Plains, New York, after the driver of a propane truck crashed into a bridge column. He had been working for two days straight with no more than five and a half hours of sleep and he fell asleep at the wheel.[29]

FACTORS LINKED TO DROWSY DRIVING

Sleep deprivation. Anyone who is sleep-deprived is at risk for driving drowsy. Those at highest risk are the following:

- Individuals twenty-five years old and younger. Their lifestyles often embrace studying into the wee hours, late-night partying, or working evening jobs, all of which lead to irregular sleep-wake schedules.
- Shift workers aged thirty to fifty-nine who have disruptive work schedules.
- The elderly, who have fragmented sleep, too little sleep, and a high incidence of sleep disorders.[30]

Alcohol. Sleepiness and alcohol are an extremely dangerous combination. The Institute for Traffic Safety reports that alcohol is involved in one third of the accidents in which the driver fell asleep. This problem is especially prevalent around the holidays when more people are drinking and more people are out on the road.[31]

According to Dr. Thomas Roth, the director of the Sleep Disorders and Research Center at Henry Ford Hospital in Detroit, "One drink on six hours of sleep is the equivalent of six drinks on eight hours of sleep."[32] Thus, even a person who has only had one drink but is sleep-deprived is a serious hazard behind the wheel of a car. He or she is stone drunk. An accident waiting to happen. It's only a matter of time.

Sleep disorders. Sleep disorders can also be a huge factor in causing accidents on the road. Sleep apnea, which causes loud snoring, disrupted sleep, and daytime sleepiness, seems to be particularly dangerous. As mentioned previously, drivers with undiagnosed sleep apnea have crash rates significantly higher than other drivers.[33] In one major trucking firm 75 percent of its drivers were found to have sleep apnea. The company physician had never heard of the disorder.[34]

Excessive work hours. Dr. Ann McCarthy of the Institute for Traffic Safety in Albany, New York, found a nearly direct relationship between hours worked and risk of falling asleep at the wheel. Of people who worked thirty-five hours or less per week, about 20 percent reported that they had fallen asleep at the wheel; of those who worked thirty-six to forty hours, 25 percent said they had dozed off; and of the people who worked fifty or more hours per week, nearly 50 percent admitted they had fallen asleep while driving.[35]

A thirty-year-old manager of a country club had been working seventy- to eighty-hour weeks. The night before his acci-

Figure 11.2. Asleep at the wheel.

dent he had a good eight-hour sleep, his first eight-hour sleep in over a year. He thought he was well rested, but in fact he was carrying a long-term heavy sleep debt. On a bright Sunday afternoon in the fall, under clear driving conditions, he fell asleep and drove under a parked eighteen-wheel tractor-trailer. The top of his car was sheared off and the engine block was driven into the passenger seat. Miraculously, he survived (see Figure 11.2).

Weight gain. A new study has suggested that being overweight may also be linked to increased traffic accidents. The findings suggest obese drivers, who may be more prone to sleep apnea, may not get enough sleep at night, which results in poor concentration and sluggishness during the day.

Late-night driving. Most crashes occur between midnight and 6 A.M., with a secondary peak between two and four in the afternoon; exactly when the body's rhythms slow down and sleepiness occurs.

Driving alone. Frequently, drivers who fall asleep at the wheel are without a companion in the car, and few if any other vehicles are on the road.[36]

Monotonous roads. Drivers are often lulled to sleep by long stretches of flat, high-speed roadways.[37] Teresa Birch, a mother of six children under the age of eight, had not had much sleep during the week prior to her accident. Halfway through her five-hour drive home to Spokane, Washington, from Seattle, she fell asleep and hit a guardrail. She was driving at 65 mph on cruise control, she was alone in the front seat, and her children were taking an afternoon nap in the back. The car was thrown into a twelve-foot-deep irrigation canal and three of her children were killed.[38]

DANGER SIGNALS

The American Automobile Association's Foundation for Traffic Safety and the Better Sleep Council warn that you could be in immediate danger of falling asleep under the following conditions:[39]

- You are feeling drowsy. Drowsiness and impairment are synonymous. Drowsiness is the last subjective event before you fall asleep.

> **DROWSINESS IS RED ALERT!**

- Your eyelids are feeling heavy.
- Your eyes burn or feel strained. They close or go out of focus by themselves.

- You have trouble keeping your head up.
- You can't stop yawning.
- You have wandering, disconnected thoughts.
- You don't remember driving the last few miles.
- You drift between lanes, tailgate, or miss traffic signs.
- You are driving at abnormal speeds.
- You keep jerking the car back into the lane.
- You have drifted off the road and narrowly missed crashing.

WHAT TO DO IF YOU BECOME DROWSY WHILE DRIVING

If you have one or more of the drowsiness symptoms listed above, **pull off the road immediately and take a nap.**

- Stop at a safe place to take a fifteen- to twenty-minute nap. Well-lighted and busy rest areas are the safest. Alternatives are an open gas station or convenience store. Avoid parking on the shoulder of the road. It is illegal except for medical emergencies and vehicular breakdowns, plus there is a risk of being hit by another car whose driver might be drowsy.
- After your nap, get some coffee or cola and engage in some physical activity like a brief brisk walk.

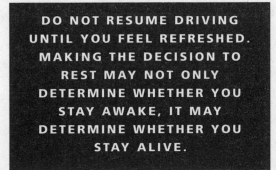

DO NOT RESUME DRIVING UNTIL YOU FEEL REFRESHED. MAKING THE DECISION TO REST MAY NOT ONLY DETERMINE WHETHER YOU STAY AWAKE, IT MAY DETERMINE WHETHER YOU STAY ALIVE.

FOR MORE INFORMATION:

The AAA's Foundation for Traffic Safety offers a free ten-page pamphlet called "Wake Up!," available through any American Automobile Association club or to those who send a stamped, self-addressed, business-size envelope to the foundation at 1140 New York Avenue NW, Suite 201, Washington, DC 20005.

The National Sleep Foundation's brochure "Drive Alert, Arrive Alive" is available by writing to the foundation at 729 Fifteenth Street NW, Fourth Floor, Washington, DC 20005, or by calling 202-347-3471.

The Better Sleep Council has literature available by writing to them at P.O. Box 19534, Alexandria, VA 22320-0534.

A DIGRESSION: STRATEGIES FOR PLANNING LONG ROAD TRIPS[40]

- Don't start a trip if you've had too little sleep, or are in the midday slump, or it is late in the day. Driving requires mental and physical alertness. Before driving long distances be sure to get a good night's sleep on a comfortable, supportive mattress.
- Plan ahead by getting plenty of sleep the week before.
- Don't plan to drive more than ten hours, the legal limit for commercial drivers.
- Try to drive during the times of day when you are normally awake, and try to stay overnight somewhere rather than driving straight through.
- Don't put yourself in a time bind. Plan for traffic congestion, bad weather, and unpredictable delays by allowing extra time for your trip.
- If you feel you *need* caffeinated drinks or over-the-counter medications just to stay awake, you may be too tired to drive. Remember, caffeine will give you a short burst of energy, but it should not be relied on as a replacement for real rest and alertness.

- Avoid long drives at night. Your body craves sleep after dark. The glare of oncoming lights increases the danger of highway hypnosis.
- If possible, don't drive alone. Sharing the driving can relieve tiredness and monotony.
- Adjust the car temperature and environment so that it's not too comfortable. Keep the temperature cool, turn the radio volume up, and avoid listening to soft, sleep-inducing music.
- Do not use cruise control. Keep your body involved with the driving.
- Watch your posture. Sitting the wrong way can easily bring on tiredness. Drive with your head up and shoulders back. Legs should be flexed at about a forty-five-degree angle.
- Take frequent breaks. Stop and get out of the car at least once every two hours, or every 100 miles or so. If you are traveling with an AAA or CAA Triptik, ask the travel counselor to indicate appropriate stops.
- Exercise during your breaks. Move your body briskly to increase your heart rate and improve alertness.
- Follow your body clock. Drive when you are normally alert and avoid driving during your body's natural "down time." Stop when you feel sleepy.
- Monitor your medications. Avoid driving if you have used medications that induce drowsiness.
- Stop for light snacks or drinks.
- Do not consume alcohol before driving. Even one drink, if you're tired, can severely impair your ability to drive a car.
- Don't let your eyes become fixed straight ahead. Scan the area from side to side, blink frequently and naturally.
- If you are driving for consecutive days, make sure you get a good night's sleep in between long drives.
- Don't drive if you have an untreated sleep disorder.

**If these antifatigue measures fail,
there is only one solution: sleep!**

CHAPTER 12

AVOIDING FAMILY
SLEEP TRAPS

Most people find it difficult enough to manage a successful career, have a satisfying social life, and still get adequate sleep. If you add to that the stresses of responding to daytime and nighttime demands of newborns, infants, children, adolescents, or elderly parents, you are likely to experience severe sleep deprivation. With some guidance, you can avoid the sleep traps that keep you (and them) from taking full advantage of the power of sleep to prepare the mind for peak performance.

TIPS FOR EXHAUSTED PARENTS OF
NEWBORNS, INFANTS, AND CHILDREN

If the only thing preventing you from sleeping like a baby is your baby, what should you do? First, take comfort in the fact that you are certainly not alone in this dilemma. On average, a parent of a new baby loses 400 to 750 hours of sleep during the first year.[1] Approximately 30 percent of young children (one to four years old) demand parental intervention at least once nightly. You can't simply ignore the problem or lock yourself in a soundproof bedroom. The best solution is to condition your child by applying many of the golden rules and sleep strategies mentioned in Chapters 5 and 6 to your child. Both you and your child will sleep more soundly and benefit from more nights of uninterrupted sleep.

Newborns (Birth to Four Months)

Minimize Nighttime Activity.
Newborns have no idea that nighttime is for sleeping; they do the majority of their resting during the daytime and are ready to play at 3 A.M. Avoid exciting the baby with singing, music, laughter, and bright light during nighttime feedings and diaper changing. Use only the daytime for playtime. It will bring about a quicker adaptation to an eventual diurnal schedule.

Eliminate Night Feedings When Appropriate.
Exhausted breastfeeding mothers of newborns should freeze their milk when possible and enlist the help of another reliable party to take over some of the night feedings. Uninterrupted

sleep is essential for good maternal care, as sleep deprivation is sure to affect your mood and stamina. As the baby grows, nighttime feedings should be eliminated through a gradual weaning process.

Infants (Five to Eight Months)

Establish a Regular Sleep Schedule.
Make sure your infant maintains a regular sleep-wake schedule, going to bed at the same time every night, including weekends. During this stage, most babies can sleep through the night, often for eight hours.

Ignore the Calls.
If you run to your infant every time he or she cries, you are being trained by your demanding child; answering only loud cries will bring more loud cries sooner. If your child wakes up often in the night crying for you, try ignoring the calls (as long as you know the infant is safe and dry). You'll find that after several (understandably unbearable) minutes the crying will subside and your child will easily fall back to sleep.

Minimize Teething Problems.
While teething can cause an infant enough pain to keep him or her awake for several nights, it should not be a cause for long-term sleep disturbances. For the occasional teething-related awakening, comfort your infant verbally as well as physically. If your child has been having sleep problems for over a week, a doctor can prescribe medicine to soothe the inflammation and provide at least temporary relief.

Toddlers (Nine to Eighteen Months)

Establish a Good Bedroom Environment.
A night-light or an open door always makes toddlers feel more secure. Keep the temperature of the bedroom around 65° F. A room that is too hot or too cold can interfere with normal sleep patterns.

Reassure and Comfort.
This is the age when separation anxiety first occurs. As a result, many children attempt to delay their bedtimes by making persistent demands for your presence. Rocking an infant creates a soothing rhythm conducive to sleep. However, it's a good idea to leave the bedroom even if your child is still awake. This way you avoid establishing a pattern of dependence that can lead to the child's having problems falling asleep later on.[2]

Provide a Surrogate Parent.
Encourage nighttime attachment to a doll or stuffed animal. Upon awakening in the middle of the night your child will feel comforted by the presence of this familiar toy.

Preschoolers (Two to Five Years)

Encourage naps.
An early-afternoon daytime nap can prevent irritability in the late afternoon when the effects of any sleep deprivation would be most evident.

"The first one to fall asleep gets today's competitive-edge award."
Drawing by Weber; © 1988 The New Yorker Magazine, Inc.

Establish bedtime rituals.

Bedtime rituals instill a feeling of security in a small child. You decide the bedtime, but let your child decide which pajamas to wear, what story to read, or what lullabies to hear.

Avoid nightmares.

Nightmares are bad dreams that seem to happen fairly often at this age. Don't let your child watch scary movies, especially those involving children; at this young age, kids still have difficulty separating fantasy from reality.

TIPS FOR EXASPERATED PARENTS OF SLEEPY ADOLESCENTS

Dealing with exhausted and moody teenagers can add stress to your life and deprive you of sleep. It's a common misconception that adolescents need less sleep than children. Teenagers actually need more sleep than prepubescent children because of the significant physiological and psychological changes that accompany the transition from childhood to adulthood.

Adolescents need approximately ten hours of sleep each night to be fully alert all day, yet they average only around six—nearly a four-hour deficit every night. It's a recipe for disaster because there is an ineluctable relationship between sleep loss and alertness. Studies at Cornell and Stanford found that only 1 percent of students report that they are fully alert all day long! Severely sleep-deprived adolescents drag themselves through high school and college like walking zombies, as if drugged by sleeping pills. They are moody, lethargic, and unprepared or unable to learn. Memory, concentration, communication skills, and critical and creative thinking are significantly affected. Education programs and your hard-earned dollars for tuition will be wasted unless your adolescent understands the power of sleep in preparing the mind for peak performance. Every teenager's potential for success will be increased if he or she follows all the relevant sleep rules and strategies detailed in Chapters 4 and 5. Here are the most essential tips for getting more A's with more zzzzzzzzzz's.[3]

> **ASLEEP IN THE COLLEGE CLASSROOM**
>
> An Ivy League college tuition is approximately $29,000 per year. If a student takes four courses per semester that meet three hours a week for fifteen weeks, that's $80 per lecture hour. If the student falls asleep in the classroom, it's an expensive hourly rate for a bedroom that's not very quiet, dark, or comfortable!

Get Adequate Sleep.

Get close to ten hours sleep a night to be fully alert during the day. Even seven or eight hours would be a vast improvement over the situation that commonly exists. Remember that good slow-wave sleep plays a significant role in the restoration of energy, in growth and development during puberty, and in immunity to viral infection. Furthermore, adequate REM sleep is mandatory for mood improvement, memory, learning, problem solving, creativity, communication skills, quick reaction time, and a host of other prerequisites for success in school, work, and personal relationships.

Establish a Regular Sleep-Wake Schedule.
Choose a reasonable bedtime that will permit adequate nocturnal sleep, then keep a regular sleep-wake schedule on school nights and weekend nights. A standardized schedule across the week will actually reduce the busy teenager's total time needed for adequate sleep.

Get Exercise to Stay Fit.
Twenty to thirty minutes of exercise at least three times a week will lead to better sleep. However, avoid exercise too close to bedtime, as this stimulates the body at a time when it should be winding down.[4]

Limit Caffeine Late in the Day.
No coffee or colas after 6 P.M. Even the small amounts of caffeine found in chocolate can disrupt nighttime sleep.

Avoid Alcohol.
Even a low amount of alcohol, if ingested by a sleep-deprived person, sets the stage for disaster in operating a motor vehicle or making responsible decisions. Exhausted adolescents, who are learning to drive and learning to drink, are most vulnerable.

EVERY TEENAGER SHOULD UNDERSTAND THE IMPORTANCE OF SLEEP. WITHOUT ADEQUATE SLEEP, THE POTENTIAL FOR SUCCESS IS SIGNIFICANTLY LOWERED. SLEEP IS JUST AS IMPORTANT AS NUTRITION AND EXERCISE IN PREPARING THE MIND AND BODY FOR PEAK PERFORMANCE.

TIPS FOR CARING FOR YOUR ELDERLY PARENTS

If you happen to take care of an elderly person, then you know that older people just don't seem to sleep very soundly. As people age they tend to spend less time sleeping and proportionately less time in REM sleep. The elderly actually need about the same amount of sleep as they did as young adults, but are unable to maintain sleep for a long block of time because of changes in the brain that accompany old age. Although many elderly people sleep less at night, they often make up for some lost sleep by napping during the day, which in turn can lead to nocturnal insomnia—a vicious cycle that's hard to break. The change in sleep patterns with age may be quite large for some people and practically nonexistent for others, but the total amount of sleep we get ideally should be the same as we enter old age.

Why Do So Many Older People Have Trouble Sleeping at Night?

Elderly people have trouble sleeping for the same reasons as do people from other age groups. But there are additional factors.

- **Chronic illnesses.** The elderly often suffer from chronic mental and physical illnesses that make it difficult to fall asleep or stay asleep. For example: An enlarged prostate is often associated with a frequent need to urinate. Dementia may be associated with confusion and heightened fear after dark (known as sundowning).

Painful conditions such as bone fractures due to osteoporosis (especially in women), arthritis, and vascular disease can make sleeping difficult. Depression, quite common in the elderly population, often causes insomnia as well.

- **Medication.** Older people have stronger reactions to medication. While people over the age of 65 make up about 13 percent of the U.S. population, they consume over 30 percent of all dispensed prescription drugs, many of which are tranquilizers and hypnotics.[5] Notable effects that these medications have on the elderly include frequent nightmares, insomnia, and hypersomnia.[6]

- **Caffeine and alcohol.** The elderly are also more vulnerable to the effects of caffeine and alcohol; that cup of tea after dinner that may not have kept a twenty-year-old up all night might do so for a seventy-year-old.[7]

- **Nursing home and hospital environment.** You would think that nursing homes and hospitals would be places where elderly people would be able to get their much-needed rest. Unfortunately, this is not the case. A study by John Schnelle of UCLA has shown that, on average, nursing home residents are exposed to thirty-two loud noises each night.[8] This includes staff members talking, sometimes even shouting, use of the intercom, doors slamming, and other patients screaming. In addition to the noise, nurses make rounds several times a night, waking up residents to give them their medication. Some residents are even woken up to take their sleeping pills!

- **Fatigue.** Many elderly people who suffer from the fatigue associated with lighter and inconsistent sleep go to bed earlier to make sure they get enough sleep. This makes the problem worse because rather than sleeping longer, they often simply wake up earlier.

- **Dwindling melatonin.** Research suggests that we produce less natural melatonin as we age.[9] Since melatonin is responsible for inducing sleepiness, it is likely that lower melatonin production in older people contributes to the change in sleeping patterns and to nocturnal insomnia.

Treatment of Insomnia in the Elderly

The medications currently being used to help the elderly sleep are mostly tranquilizers and painkillers. Because some of these drugs have undesirable side effects and the elderly are particularly susceptible to them, the threat of overmedication needs to be considered. Other methods to induce sleep with fewer side effects are being sought. One study showed that lavender oil had a mild sedative effect that brought about sleep in elderly subjects to the same extent as the sleep medication they had been taking. Subjects were also less restless during sleep.[10]

For some elderly who suffer from dementia, but are able to recognize friends and relatives, a familiar friendly person to keep them company at bedtime may help reduce some of the confusion and fear that they experience at night.[11]

Insomnia is also a part of the advanced sleep phase syndrome that often accompanies old age. (See Chapter 13, "Insomnia and Beyond.") Many elderly people wake up very early in the morning, which causes them to be tired and to nap during the afternoon. Assuming the change in sleeping pattern is a result of natural aging and is not due to other problems associated with aging, there are some remedies:

- **Chronotherapy.** In chronotherapy, bedtime is set later every few days until the desired bedtime is reached, thereby resetting the internal biological clock and its circadian rhythm.
- **Melatonin therapy.** A recent study showed that melatonin therapy helped bring sleeping patterns in the elderly closer to those of younger people.[12] In this study, melatonin was adminis-

tered in the form of a special controlled-release formula. It is important to consult a physician before taking melatonin.

Strategies to Help the Elderly Sleep Better

Fifty percent of adults over age sixty-five claim to get irregular sleep, although experts point to studies that indicate that the number might be as high as 90 percent. Senior citizens should follow all the rules and strategies set forth in Chapters 4 and 5, and also pay special attention to the following:

Avoid Caffeine.

Restrict caffeine intake after 6 P.M. As we get older, coffee, tea, soda, and chocolate may make it more difficult to fall asleep.

Monitor Prescriptions.

Consult your physician about the sleep side effects of medications you are taking. Some medicines cause drowsiness, and others cause sleeplessness.

Avoid Over-the-Counter Sleeping Remedies.

Stay away from over-the-counter sleep medications. Although they can make it easier to fall asleep, they often cause light and fragmented sleep, as well as early-morning waking up.[13]

Get Outdoors During the Day.

Studies have shown that exposure to natural light results in better nighttime sleep.[14]

Stay Physically Active.

A late-afternoon or early-evening stroll will help you sleep more deeply at night.

Stay Mentally Active.

Engage in hobbies, academic courses, volunteer work, and social activities that will keep you mentally alert in the daytime and early evening.

Limit Naps.

If you feel the need to take a daytime nap don't sleep longer than twenty minutes. Longer naps make it more difficult to sleep at night.

A DIGRESSION: SLEEP AND GENDER

There are some fascinating differences in the sleep habits of men and women.

- Women sleep four minutes less than men on workdays and fourteen minutes less on days off.[15] Women sleep forty to fifty minutes less than men if they cater to the needs of a baby or toddler.
- Scientists have discovered that there are differences in the circadian rhythms of males and females. The fraction of sleep is larger in women. The temperature rhythms are identical in males and females; the sleep-wake rhythms are significantly shorter in females.[16] "Women have shorter periods of deep (non-REM) sleep than men, making them more vulnerable to 'tossing and turning' bed partners and fitful babies."[17]
- Women are more likely than men to suffer from insomnia.[18]
- Young adult women report insufficient sleep, excessive daytime sleepiness, difficulties in maintaining sleep, and the absence of feeling refreshed in the morning more frequently than do young adult men.[19]
- More elderly women than men suffer from sleep problems.[20]
- Sixty-six percent of night eaters are women.[21]

- Male smokers tend to have disturbing dreams; female smokers experience excessive daytime sleepiness.[22]
- Women tend to have significantly more nightmares than men.[23]
- More friendly acts occur in women's dreams than in men's dreams.[24]
- Ninety percent of people suffering from REM sleep behavior disorder are male.[25]
- Chronic snoring is found in 24.1 percent of men and in 13.8 percent of women.[26]
- Women are more likely than men to talk in their sleep.[27]
- During REM sleep, men have penile erections. This is one way that doctors can investigate physiological impotence. Younger men have between four and five each night, and older men in their seventies have about two or three. Occasionally this sexual arousal can lead to orgasm. Hence, the term "wet dream."
- A similar phenomenon occurs in females. During REM sleep, women experience clitoral engorgement and increased vaginal lubrication. In fact, it is also possible for women to have an orgasm while sleeping. Primary orgasmic dysfunction is the term used to describe women who have only experienced orgasm during sleep or fantasy.[28]
- The sexual content present in the dreams of a woman can actually reflect the stages of her menstrual cycle. For example, sexuality is strong during the postovulation stage and is at a maximum during menstruation. Sexual drive is lowest during the postmenses phase. Hostility is maximal in the postovulation stage and at a low during menstruation. The content of dreams is more sexual or hostile in nature during the times when these waking emotions are at a high. During the postmenses phase, dreams are rarely sexual in content.[29]
- Women sleep much less in the premenstrual phase of the cycle and often feel more tired, irritable, confused, and depressed during this phase.[30]
- During pregnancy sleep is less continuous, especially during the last trimester.

- Although age in itself does not affect sleep patterns, studies have shown that women who experience hot flashes get worse nighttime sleep than those who do not.[31]
- Sleep loss associated with menopause has been shown to cause irritability and depression.[32]

PART V

WHEN TO CALL THE SLEEP DOCTOR

Sleep's the only medicine that gives ease.

—SOPHOCLES, *Philoctetes*

SLEEP

CHAPTER 13

INSOMNIA AND BEYOND

I f you're having trouble sleeping, you've got company. In a 1995 national survey, the Gallup Organization reported that 49 percent of adults say they have problems with their sleep.[1] On any given night, millions of us have insomnia—we either cannot get to sleep or cannot maintain sleep. When we battle with the night, our sleep is nonrefreshing and nonrestorative. The daytime consequences range from drowsiness to decreased productivity, increased accident rates, health problems, and a reduction in the quality of life. In fact, fatigue is the most common complaint heard by physicians.

In 1996, Dr. Thomas Roth, director of the Sleep Disorders and

Research Center at Henry Ford Hospital in Detroit, estimated the total annual economic costs attributable to insomnia, including decreased productivity at work and health and property costs related to accidents, to be in the range of $30 billion to $35 billion. This does not include costs difficult to calculate, such as those related to possible increased mortality, academic failure, and lost job opportunities.[2]

The economic figures hardly reflect the magnitude of our problems with sleep. Many sleep disorders go unreported because the troubled person is not awake or alert when the symptoms appear. These individuals are chronically exhausted but do not know why, and may be regarded by others as simply lazy.

Sometimes the symptoms of sleep disorders can be quite bizarre—practically unbelievable—and sufferers may not understand their own actions. They may be afraid of the stigma associated with their behavior, or they may fear the revelation of underlying psychiatric illness and therefore avoid admitting to others, including those who can help, that they have a problem. Consider the following real-life experiences taken from the files of the American Sleep Disorders Association:

- The sleeping woman who awoke to find herself in a grocery store aisle wheeling a cart filled with fifty-six boxes of cornflakes.
- The woman who drew a map of the United States on her bedroom wall and then filled in the capitals of every state—all in her sleep.
- The air traffic controller whose only way to keep from falling asleep on the job was to stand while working.
- The woman who gained over forty pounds eating chocolates while she slept.
- The man in the back of a moving trailer who sleepwalked out the door and was killed on the highway.

- The sleeping woman who drove to the airport, bought a plane ticket, and flew halfway to California before waking up.

Some of these anecdotes may appear humorous at first glance, but people with sleep disorders really do suffer, as might innocent others. Imagine being on the same road with someone driving in his sleep, or being a passenger in a Boeing 747 on final approach, your fate in the hands of a jet-fatigued pilot or an air traffic controller who is barely awake. Or imagine sharing a bed with someone who snores loudly, or who kicks, punches, and thrashes hundreds of times each night. Hardly restful for either of you. Think of that ninety-year-old California man who strangled his wife of sixty-two years to death because she kept him awake at night with her persistent cough.[3]

Because sleep deprivation and fatigue are so pervasive in our stress-filled, hectic society, many physicians consider lack of alertness, low energy levels, and moodiness to be within the normal range of functioning. They sometimes dismiss such commonplace complaints as not being symptomatic of a legitimate medical illness.

Even when a patient complains of falling asleep at inappropriate times during the day, or describes rather bizarre nocturnal behaviors, a sleep disorder often remains undiagnosed and untreated.[4] As recently as 1985, Dr. William Dement reported, "There is not one single medical school that devotes systematic and valid attention to sleep disorders."[5] Historically, physicians have been ill prepared to recognize sleep disturbances, or even ask sleep-related questions when taking a patient's medical history. The National Commission on Sleep Disorders Research searched a primary-care-practice database of 10 million patients expecting to find the estimated 100,000 cases of a common and life-threatening sleep disorder, obstructive sleep apnea. They found only seventy-three instances where the diagnosis was made.[6]

Sleep medicine, established as a recognized medical specialty in 1996, is still in its infancy. Yet giant strides are being made. Medical schools and continuing medical education programs are beginning to offer training in sleep disorders medicine. The number of sleep disorders centers in the United States has increased from a handful in the early 1980s to 1,500 as of 1997. Three hundred twenty-five of the clinics have received voluntary accreditation from the American Sleep Disorders Association.

Might you have a sleep disorder that requires treatment from a health-care professional or sleep center? If you answered true to any of the items on Self-test D in Chapter 2, or if any of the disorders described in the rest of this chapter describe what you experience, consider seeking the help of a sleep specialist at one of the accredited sleep disorders clinics listed in Appendix D. Doing so could vastly improve your quality of life, or even save your life.

In this chapter you will find a description of the most common sleep disorders, but the list is not exhaustive. For a detailed classification and description of all known sleep disorders, consult *The International Classification of Sleep Disorders: Diagnostic and Coding Manual*. It is available from the American Sleep Disorders Association, 604 Second Street SW, Rochester, MN 55902.

DYSSOMNIAS

The dyssomnias are disorders of sleep that produce difficulty in initiating or maintaining sleep, or produce excessive sleepiness. They include insomnia, sleep apnea, narcolepsy, restless legs syndrome, periodic limb movement disorder, hypersomnia, and delayed or advanced sleep phase syndrome.

Insomnia

If you have trouble falling asleep, or if you wake up a number of times at night, or if you wake up earlier than you would like, or if you always feel as though you need a nap during the day, or if you fall asleep while watching TV or while waiting for a traffic light to change, you might very well have insomnia. If so, you're not alone. One in every two Americans has experienced insomnia, the most common sleep disorder.[7]

Those with insomnia experience an insufficient amount of sleep, or do not feel rested after having slept. This often impairs social and occupational functioning, and can be accompanied by feelings of restlessness, irritability, anxiety, daytime fatigue, and tiredness.[8]

Transient or short-term insomnia lasts no more than a few nights to two or three weeks. It can be caused by any of the following:

- Temporary illness
- A change in the timing of sleep (for example, jet lag, or rotating shift-work schedules)
- Environmental factors such as excessive noise and light, or an uncomfortable bedroom temperature
- A traumatic experience, such as a death in the family[9]
- Excitement regarding an upcoming or recently occurring event
- A stressful problem

Insomnia is not a single syndrome with a single solution for all situations and causes. Transient or short-term insomnia is best treated by practicing the golden rules of sleep and the sleep strategies described in Chapters 5 and 6. Following the suggested procedures and filling out the sleep logs for six weeks should help solve most minor sleep problems.

Forty percent of people who suffer from insomnia try to self-medicate with over-the-counter drugs and other proposed remedies such as melatonin and L-tryptophan.[10] This is potentially dangerous if not done under the supervision of a knowledgeable physician; information regarding the safety and effectiveness of many over-the-counter sleep enhancers is limited.[11] If warranted, sedative-hypnotic drugs may be prescribed by a doctor for a short period.

Chronic insomnia is by definition insomnia that lasts more than two or three weeks. It is often related to underlying medical conditions such as dementia, Parkinson's disease, sleep-related epilepsy, and cerebral degenerative disorders or psychiatric problems such as anxiety, mood and panic disorders, psychoses, and alcoholism. Chronic insomnia should be treated by a physician well-versed in sleep medicine.

For further information on insomnia, contact the National Sleep Foundation, 729 Fifteenth Street NW, Fourth Floor, Washington, DC 20005. Tel: 202-347-3472.

Obstructive Sleep Apnea

Are you a loud snorer who disturbs your bedpartner? Do you feel tired upon awakening? Have you been known to hold your breath, choke, or gasp during sleep? If so, you might very well have sleep apnea.

Obstructive sleep apnea occurs because of a sleep-induced failure of the throat muscles to hold the airway open against the suction created by efforts to breathe.[12] It is characterized by repetitive episodes of upper airway obstruction that occur during sleep, causing cessation of breathing for thirty to ninety seconds at a time, up to 600 times each night. Alternating with these long pauses in breathing are loud snores, or brief gasps, and often whole-body movements. The snoring is so loud that it will likely disturb the sleep of a bed partner or anyone else sleeping in close proximity.

The individual with sleep apnea must awaken momentarily to resume breathing each time there is an airway obstruction. The unsuspecting person might think he has slept the whole night when in fact he may have woken up hundreds of times each night. Sufferers of obstructive sleep apnea are therefore extremely sleep-deprived and often complain of excessive day-time sleepiness. The overwhelming tendency to fall asleep becomes evident when the person is in a relaxing situation, such as sitting in a comfortable chair reading, or watching television. He will often fall asleep in boring meetings, or while attending movies or concerts—even exciting athletic events. Hopefully, he will not be driving a car when the sleepiness manifests itself. Imagine how tired you would be if you had to wake up 600 times a night to resume breathing!

A forty-five-year old male was referred to a sleep clinic for evaluation and treatment. He had a history of loud snoring with repetitive pauses in breathing, and daytime sleepiness. For the past year, the patient and his wife had not slept in the same bedroom because of his severe snoring and the wife's inability to sleep. The patient's sex drive had decreased over this period of time, and he was overweight. He woke up in the morning feeling unrefreshed, and had severe problems maintaining daytime alertness, even at work. He fought to stay awake in business meetings or one-on-one conversations. He would drift off, wake up, and realize he was uncertain of what the conversation was about. He had difficulty driving to work. The first thing in the morning, on a drive as short as three miles, he would find himself nodding off. He also fell asleep reading, or while watching television. Recently, he had been sleeping in a chair in an attempt to improve his breathing episodes, but this did not help. His wife described a typical cyclic breathing pattern of loud snoring, cessation of breathing, a loud snort, and resumption of breathing.[13]

This patient was evaluated by an all-night sleep recording. He was diagnosed as having severe and potentially life-threatening sleep apnea—certainly life-threatening in terms of a potential motor vehicle accident. Twenty-four percent of sleep apnea sufferers reported falling asleep at least once a week while driving their cars.[14]

Obstructive sleep apnea is most common in middle-aged men. Overweight people are at higher risk, and men are eight times more likely than women to develop it.[15] In the United States, at least one out of every 200 people suffers from obstructive sleep apnea, but 95 percent don't even know it.[16, 17] (Central sleep apnea is not as common as obstructive sleep apnea, but is also a dangerous condition. In this form of apnea, the neural communication between the lungs and the brain appears to malfunction, delaying the signal to breathe.)[18]

Because the health risks of apnea are so serious, it is necessary to avoid alcohol and sleeping pills, which could have fatal consequences. Sometimes apnea can be cured through weight loss, or by sleeping on one's side.

Mild sleep apnea is often treated with the drug Vivactil, or with an apparatus made to ease breathing during sleep. The most commonly used apparatus is called a Continued Positive Airway Pressure (CPAP) machine, and consists of an air pump and a mask that is worn over the nose and mouth at night. A low-pressure stream of air keeps the throat's airway passage open.[19]

More serious cases of sleep apnea are often treated surgically, by repositioning the jaw to make the airway passage larger; removing or shrinking excess tissue in the throat through laser

A NOTE ON SNORING

It is important to note that not all snoring is a sign of apnea. Only about 1 in 100 snorers suffers from sleep apnea, yet between 30 percent and 40 percent of adults snore.[25] According to the experts at Stanford University's Sleep Well website, "Snoring is a noise produced when an individual breathes (usually produced when breathing in) during sleep which in turn causes vibration of the soft palate and uvula (that thing that hangs down in the back of the throat).

"The word 'apnea' means the absence of breathing, where the airway is completely blocked for a period of time, usually ten seconds or longer. This silence is usually followed by snorts and gasps as the individual fights to take a breath. The individual snores so loudly that it disturbs others. Obstructive sleep apnea is almost certain to be present."[26]

Primary snoring (snoring that is not due to apnea) is not life threatening, as sleep apnea is, and does not result in chronic fatigue. It may, however, lead to insomnia for a bed partner. The sound of snoring can reach a thunderous ninety decibels, which exceeds the government standard for noise in the workplace.

If you snore and you sleep on your back, try switching to sleeping on your side instead.[27] You can sew a pocket into the back of your bedclothes and put tennis balls or golf balls inside. This will discourage you from sleeping on your back. There are also mouth devices and surgical procedures that will alleviate snoring.

The important point is that sleep apnea must be absolutely ruled out as the reason for snoring before any other preventative techniques are used. If you snore heavily, consult your physician, and insist that a proper test for sleep apnea be done.

or radio-wave surgery; or by means of a procedure called a tracheostomy, in which a valve is surgically implanted beneath the blockage in the upper airway passage so incoming air can bypass it.[20] The valve can be closed during the day so the person can breathe and speak normally. Apnea in children can often be treated by the removal of their tonsils or nasal polyps.[21]

People with sleep apnea are twice as likely to have a heart attack or stroke within ten years as those without it.[22] Heartburn, cardiac arrhythmia, mild hypertension, and secondary depression, anxiety, and irritability have been observed in some patients with obstructive sleep apnea.[23] Recent research suggests that apnea may actually cause heart disease, and people with coronary artery disease and apnea may be at risk for nocturnal sudden death.[24]

For further information, contact the American Sleep Apnea Association, 2025 Pennsylvania Avenue NW, Suite 905, Washington, DC 20006. Tel: 202-293-3650. You also might want to check a good sleep apnea resource website: http://www.bway.net/~marlene/awchapt.html.

Narcolepsy

Are you always tired, no matter how much sleep you've had? Do you ever collapse after hearing a funny joke, or after becoming very angry? Do you have any relatives who are always sleepy? If you answer yes to these questions, you might have narcolepsy. Narcolepsy is characterized by excessive daytime sleepiness, cataplexy, sleep paralysis, and hypnagogic hallucinations (explained below). Narcolepsy is thought to be an attack of REM sleep during the day.

A person with narcolepsy will have recurring episodes of naps, or lapses into sleep. The narcoleptic will sleep for fifteen to twenty minutes and awaken refreshed, but within the next two to three hours will begin to feel sleepy again. This pattern repeats itself throughout the day. Sudden sleep attacks can occur in situations when it is quite inappropriate to sleep, such as while eating, conversing, or driving.[28]

A primary symptom of narcolepsy is cataplexy, a momentary

loss in muscle tonus provoked by strong emotion, such as surprise, laughter, anger, or elation. The patient may experience "a mild sensation of weakness with head droop, facial sagging, jaw drop, slurred speech, and buckling of the knees to complete postural collapse, with a fall to the ground. The duration is short, ranging from a few seconds to several minutes, and recovery is immediate and complete."[29] Someone who has narcolepsy may collapse into REM sleep after hearing a joke, or after hitting a good golf shot. REM sleep, lasting from several seconds to thirty minutes or more, sometimes follows a cataplectic attack.[30]

The narcoleptic often experiences sleep paralysis, an inability to move or speak during the transition between sleep and wakefulness. This is limited in duration to a minute or so, but can be frightening, especially when accompanied by a sensation of inability to breathe, and by hypnagogic hallucinations, which are vivid perceptual experiences at sleep onset. Patients hallucinate the presence of someone or something, and feel fear or dread. "Hallucinatory experiences, including being caught in a fire, drowning in a bathtub, being attacked, or flying through the air, are frequently reported."[31]

A thirty-four-year-old female was referred for evaluation of inappropriate daytime sleepiness. She reported sleeping seven and a half hours at night, then another one and a half to two hours in the morning. She also naps for one to two hours in the afternoon. In spite of this, she is likely to be sleepy on drives as short as ten minutes. She often will have to pull over to the side of the road because of an overwhelming feeling of sleepiness. She has fallen asleep reading, watching TV, or talking to other individuals. She has fallen asleep while standing in a grocery store checkout line, and at work. When she laughs or gets angry, her legs buckle beneath her. She reports having hypnagogic hallucinations. When she is falling asleep she has a sensation that someone is in the room with her. Occasionally this is a tactile sensation.[32]

The patient described above was evaluated by an all-night sleep recording and a daytime nap study. Like most narcoleptic patients, she had sleep-onset REM periods, going directly from wakefulness into REM sleep without prior slow-wave sleep. She was diagnosed with narcolepsy and was instructed not to drive because of her pathologic level of daytime sleepiness until her sleepiness was brought under control.

Narcolepsy usually begins in late childhood or adolescence. Excessive daytime sleepiness is almost always the first symptom to appear.[33] Patients have difficulty in school, or at work, because they are constantly falling asleep. Not surprisingly, narcolepsy impairs social functioning and often leads to marital disharmony. According to the National Sleep Foundation, about 250,000 Americans, or 1 in 2,000, suffer from narcolepsy, about the same incidence as that of multiple sclerosis.[34] Yet 80 percent of those with narcolepsy are undiagnosed. Narcolepsy is a genetically linked disorder. People with a relative who has narcolepsy have a sixty times higher risk for the disorder. If one parent has narcolepsy, the chances are 1 in 20 that a child will have the disorder.

The symptoms of narcolepsy can be lessened by taking a ten- to twenty-minute nap every two hours throughout the day, by avoiding alcohol and sleeping pills, and by getting adequate sleep on a consistent basis. Though taking frequent naps is not an ideal solution, it's better than running the high risk of a spontaneous sleep attack. Narcolepsy is typically treated by using appropriate medication: tricyclic antidepressants for reducing cataplexy, and stimulants to counter the daytime sleepiness. Cataplexy, sleep paralysis, and hypnagogic hallucinations are likely to decrease over time, but excessive daytime sleepiness is usually a lifelong problem. But with proper medical attention narcolepsy can be managed so that the patient is able to lead a satisfactory life.

For more information, contact the Narcolepsy Network, P.O. Box 42460, 10921 Reed Hartman Highway, Cincinnati, OH 45242; website: http:www-leland.stanford.edu/~dement/narco. html##nn.

Good websites to search for further information on narcolepsy are the following:

Stanford University Center for Narcolepsy
http://www.hia.com/hia/narcoctr/

Narcoleptics and Partners (N.A.P.)
http://www.uic.edu/depts/cnr/info.html

Young Americans with Narcolepsy (Y.A.W.N.)
http://www.hep.umn.edu/~tpi-web/YAWN/YAWN.html

Restless Legs Syndrome

Do your legs ache, tingle, or itch just before you fall asleep? Does your bed partner complain of being kicked during the night? If so, you need to consult with your physician. Restless legs syndrome is a sleep disorder "characterized by disagreeable leg sensations, usually prior to sleep onset, that cause an almost irresistible urge to move the legs."[35] The sensations may be pain, discomfort, itching, pulling, tingling, or prickling. If the patient moves his legs, there is partial or complete relief of the sensation, but the symptoms return upon cessation of the leg movements. Having restless legs syndrome often leads to periodic involuntary limb movements both during sleep and during the day.[36] It is often accompanied by intense anxiety or depression.

A sixty-two-year-old female was referred for evaluation of difficulty in falling asleep, and repetitive leg movements, symptoms she has had for fifteen to twenty years. She describes a creepy, crawly sensation in her legs and feet, with the sensation that she must move her legs. This tends to come on when she is lying down to go to sleep. For several nights each week it takes her more than thirty minutes to fall asleep. When she develops the crawly sensation in her legs and the feeling that she needs to move about, she will get up and go to another room to read or do other activities. She dreads going to bed because she is worried she will have difficulty falling asleep. On those nights when she sleeps poorly, she feels she may lie awake in bed or read as late as four to five in the morning before she will fall asleep. Her anxiety about not being able to fall asleep has become a vicious cycle. The more anxious she gets, the more it prolongs the period of time before she is able to initiate sleep; consequently, if she lies in bed long enough, she will develop more symptoms of restless leg syndrome.[37]

Between 5 and 15 percent of the population suffer from restless legs syndrome.[38] The peak onset period is usually middle age, and it may be seen for the first time in advanced old age. It is more common in females. Pregnant women, people who suffer from rheumatoid arthritis, and people who are anemic are at higher risk.[39]

Restless legs syndrome is usually treated with prescribed medication, such as Sinemet, Parlodel or Klonopin, and relaxation therapy.[40]

For more information, contact the Restless Legs Syndrome Foundation, 303 Glenwood Avenue, Raleigh, NC 27603. For a list of regional associations, see the foundation's website: http://www.rls.org/sguslst.html.

Periodic Limb Movement Disorder

Does your bed partner say that your legs jerk during the night? Is your sleep unrefreshing? Periodic limb movement disorder is "characterized by periodic episodes of repetitive and highly stereotyped limb movements that occur during sleep. The movements typically occur in the patient's legs, and consist of an extension of the big toe in combination with partial flexion of the ankle, knee, and sometimes hips."[41] This often results in partial arousals from sleep. However, as with sleep apnea, the person remains unaware of the many nocturnal disruptions and does not know why he or she is so tired during the day. The limb movements often disrupt the sleep of the bed partner.

Periodic limb movement disorder may accompany obstructive sleep apnea, narcolepsy, and insomnia, and is quite common in people who suffer from restless legs syndrome as well. Around 34 percent of people over the age of sixty have periodic limb movement disorder. The medical treatment for periodic limb movement syndrome is similar to that for restless legs syndrome.

Hypersomnia

Are you sleeping almost all day? Eating excessively? Hypersomnia implies too much sleep, either in the form of prolonged nocturnal sleep or excessive daytime sleepiness. There are three varieties of the disorder:

Recurrent Hypersomnia (Including Kleine-Levin Syndrome)

This disorder involves periodic episodes of hypersomnia lasting up to several weeks. The excessive sleepiness may or may not be accompanied by binge eating and hypersexuality.[42]

Idiopathic (i.e., of unknown cause) Hypersomnia

Idiopathic hypersomnia is similar to narcolepsy, but does not include cataplexy. Excessive daytime sleepiness and sleep onset at inappropriate times are characteristic of this disorder.[43]

Posttraumatic Hypersomnia

Posttraumatic hypersomnia occurs as a result of head injury and is usually seen with other symptoms of head trauma such as headaches, difficulty concentrating, and memory impairment. Posttraumatic hypersomnia typically begins immediately after the trauma, but in some instances onset can be delayed for six to eighteen months.[44]

"Healthy" Hypersomnia

Some people just sleep a lot. "Long sleepers" have what is known as "healthy" hypersomnia.[45] They simply need more sleep to feel well rested (typically more than ten hours a day) than the average person their age. "Healthy" hypersomnia is a real problem only if your sleeping pattern conflicts with your daily schedule.

Just as there are "long sleepers" who need more sleep than is considered normal, there are also "short sleepers" who average less than five hours of sleep each night but still remain fully rested.[46] Most people aren't so lucky. Both long and short sleepers make up a very small percentage of the population, and there is nothing you can do to make yourself need more or less sleep. The level of sleep you require to feel fully rested is determined by your genes.

Delayed/Advanced Sleep Phase Syndrome

Do you find it difficult to fall asleep until well after midnight, yet are able to sleep a normal length of time if you don't have to get up for work? Or, do you fall asleep too early in the evening and wake up too early in the morning? If so, your biological clock may be out of sync with your preferred sleep-wake schedule.

Someone with delayed sleep phase syndrome has sleep-onset and wake times that are later than desired, but little or no difficulty maintaining sleep once it has begun. Many adolescents, because of their neurophysiology as well as school and social schedules, suffer from this syndrome (see Chapter 12, "Avoiding Family Sleep Traps"). Teens often do not want to sacrifice their social lives or their after-school jobs for the sake of a full night's sleep. They do not get to bed until late at night and then have to wake up for early-morning classes. The result is that they are "walking zombies" for the first few hours of the day. Workers on rotating shifts, college students who stay up late to study, and eastbound jet-lagged travelers can also experience delayed sleep phase syndrome. Effects can be minimized by avoiding alcohol, sedatives, or stimulants, and by practicing good sleep hygiene, including maintaining consistent sleep-wake schedules.

Advanced sleep phase syndrome is the opposite of delayed sleep phase syndrome. Sufferers of advanced sleep phase syndrome fall asleep too early in the evening and wake up too early in the morning, often before dawn. Like delayed sleep phase syndrome, advanced sleep phase syndrome becomes a problem when it conflicts with work and social schedules.

The sleep phase syndromes are most effectively treated with bright light (2,000 to 10,000 lux) therapy. (Lux is a measure of illumination. The intensity of office lighting is about 200 to 500 lux. A bright overcast sky provides about 10,000 lux.) Those with delayed sleep phase syndrome are treated with exposure to daylight first thing in the morning, and are asked to avoid bright light in the evening. People with advanced sleep phase syndrome are exposed to daylight full-spectrum light (such as Vita-Lite or Growlux light—the same lamps that people use to grow plants!) during early evening hours. They must keep their bedroom dark during early morning hours. To establish the proper treatment schedule and to avoid damage to the eyes, consult a sleep specialist before you attempt any bright light therapy.

PARASOMNIAS

The parasomnias are a group of disorders that are not associated with the processes of sleep per se, but rather are physical abnormalities that occur for the most part during sleep. They include REM sleep behavior disorder, sleep terrors, sleepwalking, bruxism and enuresis, and sudden infant death syndrome.

REM Sleep Behavior Disorder

Do you physically act out your dreams? Injure yourself and/or your bed partner? Fly out of bed and have frightening dreams? People with REM sleep behavior disorder actually attempt to act out their dreams. They kick, punch, leap, and run from bed, often injuring themselves and/or their bed partners. One case in England resulted in a man shooting his new bride to death while he was dreaming of being pursued by gangsters. Here are two other cases:

A sixty-one-year-old male was referred for evaluation of nocturnal dream episodes which were being acted out. He is a retired police officer. His dreams typically involve executing an arrest, or responding to some sort of emergency against an attacker. It might involve handcuffing, or trying to physically restrain, an assailant. The patient's wife has been subjected to physical injury related to the acting out of the repetitive dream sequences. On one occasion the patient attempted to climb out the bedroom window, and on several other occasions he had dreams of standing at a urinal and in fact did urinate in bed. He has episodes of calling out or yelling as part of his dream sequences. The patient goes to bed around 11 p.m. and the sequences occur around 1:15 a.m., 4 a.m., and later in the morning. They happen two to four times per week and are always violent in nature.[47]

A seventy-six-year-old woman was referred for evaluation, having reported three dream-related abnormal nocturnal behavior episodes. The first episode involved a dream that she was being chased by a train. She woke up after running into her bedroom wall and injuring her face. In the second episode, she dreamed she was calling her son on the phone. She awoke to find that she had actually gotten up and made the telephone call. Her final episode was the result of a dream that there were burglars in her home. In response, she jumped out of bed and knocked over furniture and a lamp trying to escape. At first she sought psychiatric help because she understandably thought she was going insane. She was extremely frightened and feared she might severely injure herself and possibly fall out of her bedroom window.[48]

We usually can't act out our dreams. During REM sleep a part of our brain keeps us from moving our arms and legs, although we can still breathe and move our eyes. REM sleep, in essence, is characterized by a highly active, dreaming, brain in a "paralyzed" body. When the normal movement-inhibiting mechanism fails, some people, usually men over the age of sixty, may develop REM sleep behavior disorder and be able to act out their dreams. The risk of developing REM sleep behavior disorder increases with age, and men are more likely than women to develop it.

REM sleep behavior disorder can be treated with medication, such as Klonopin.[49] Patients are advised to make the sleeping environment as safe as possible, locking windows and removing sharp or fragile objects from the bedroom.

Sleep Terror

Do you wake up terrified in the middle of the night, and not know why? Sleep terror is a sudden arousal from slow-wave sleep with a piercing scream or cry, accompanied by profuse sweating and intense fear. It's not a dream or nightmare, but rather like having a panic attack in your sleep. Sleep terror is especially frightening, because unlike nightmares, it occurs without dream recall. People experiencing sleep terror do not know how they became so frightened.

Because sleep terror does not occur during REM sleep, movement is not inhibited. Concurrent sleepwalking is not uncommon. People with sleep terror have seriously injured themselves and their bed partners. One man was accused of brutally attacking two elderly people in a different house from the one in which he had fallen asleep. He had complete amnesia for the event, and it is postulated that he committed the act during an episode of sleep terror.[50]

In susceptible people, an episode of sleep terror can be induced by stress, sleep deprivation, or sleeping in a different bed.[51] Sleep terror occurs predominantly in childhood, between the ages of four and twelve, and often ceases by adolescence. An estimated 15 percent of children and adolescents experience the disorder. It's more common in males, runs in families, and can be attributed to a nervous system that has yet to mature.

Sleep terror is typically treated with medication, frequent exercise, and adequate sleep. Consult your physician if you or your child experiences this frightening but treatable problem.

Sleepwalking (Somnambulism)

Do you walk in your sleep? Sleepwalking (somnambulsim) is initiated in slow-wave sleep, and can range from sitting up in bed to walking and even to frantic attempts to "escape."[52] Sleepwalkers do not appear to be sleeping, but they are deeply asleep. Their eyes are typically wide open, with dilated pupils.[53] Sleeptalking can also be observed during sleepwalking episodes. The episodes may or may not be associated with sleep terror.

People who are sleepwalking are not acting out their dreams, but manage to do some pretty incredible things. Some sleepwalkers have gone on shopping sprees; others have traveled long distances on airplanes before waking up. One woman made herself a meal of buttered cigarettes and cat food sandwiches.[54] Some midnight snack!

Do not try to awaken a sleepwalker unless it is an emergency. First of all, sleepwalkers are not easily woken up. One man who frequently sleepwalked tied himself to his bed frame, believing the tug from the rope would wake him up in time to prevent a sleepwalking episode. Instead, he ended up sleepwalking anyway—and dragging his bed behind him![55] Also, attempting to awaken a sleepwalker may result in the sleepwalker's attacking you. The best solution is to gently guide the sleepwalker back to bed or to another place to lie down without waking the person up.

Sleepwalking can be triggered by fever, sleep deprivation, or emotional upset.[56] Approximately 10 to 15 percent of children five to twelve years old have at least one episode of sleepwalking.[57] The episodes last from thirty seconds to thirty minutes and usually do not occur more than once a night. Sleepwalking can result in physical harm to the sleepwalker or someone around him or her. Falls and injuries from walking into walls and furniture are common, so parents should carefully arrange for a safe sleeping environment. More seriously, there are a few cases that have resulted in homicide or suicide.

Sleepwalking is a genetically linked disorder (it tends to run in families) and is thought to be the result of the incomplete development of part of the brain. Sleepwalking usually ends spontaneously after adolescence—when our brains have fully matured. Thus, usually no medical treatment is advised for treating youngsters, just keeping a safe environment.

Tooth-grinding (Bruxism)

Eighty-five to 90 percent of us occasionally grind or clench our teeth during our sleep. A smaller portion, about 5 percent, do it chronically. Tooth-grinding (bruxism) occurs in about half of all normal infants and is part of the teething process.[58] Frequent tooth-grinding in adults, often caused by stress, can lead to dental damage and injury, and facial pain associated with temporomandibular joint (TMJ) disorders.[59] It's also unpleasant for the bed partner to hear the sounds made by the grinding teeth.

There are a number of ways to prevent bruxism, including the use of a plastic tooth guard, or a splint, designed by a dentist. Relaxation techniques have also proved helpful. Check with your dentist if you grind or clench your teeth during sleep.

Bed-wetting (Sleep Enuresis)

Persistent bed-wetting, or sleep enuresis, is considered a disorder after the age of five. It occurs in all sleep stages, and daytime bladder control can be normal. While the prevalence of bed-wetting in childhood decreases with age, about 3 percent of adolescents between the ages of twelve and eighteen continue to wet their beds.[60] Bed-wetting has a hereditary component. Approximately 77 percent of children whose parents both wet their beds as children are bed wetters themselves.[61] A congenitally small bladder, bladder infections, allergies, obstructive sleep apnea, or metabolic or endocrinologic disorders may be predisposing factors. Contrary to popular belief, bed-wetting is almost never emotionally or psychologically caused; less than 1 percent of bed-wetting has an emotional source.[62]

There are devices on the market that use moisture-sensitive mattress covers or underwear with alarms to condition the child over a period of time to wake up in response to a full bladder. If the bed-wetting is due to medical or psychological problems, it may disappear with the successful treatment of the causes. In some instances, medication such as imipramine or desmopressin acetate will be prescribed.[63]

Sudden Infant Death Syndrome (SIDS)

Sudden infant death syndrome (SIDS), the number one cause of neonatal and infant death in the nation, is arguably the most tragic of the sleep disorders. Occurring in 1 or 2 of every 1,000 live births, SIDS is characterized by the unexpected sudden death during assumed sleep of otherwise apparently healthy infants. It has not been unequivocally demonstrated whether the primary cause of death is respiratory or cardiac failure.[64]

Risk of SIDS is highest between ten and twelve weeks of age. Unfortunately there is no definite way to predict the possible onset of the disorder. Some babies are at particular risk:

- Infants born with low birth weight are five to ten times as likely to die of SIDS.
- Twins and triplets, even at normal birth weight, are twice as likely to die of SIDS, and after one twin dies, the surviving twin also has an increased chance of dying from SIDS.
- Babies whose previous siblings died of SIDS are two to four times as likely to die of it.

- Six percent of infants with infant sleep apnea die of SIDS.
- Black and Eskimo infants are four to six times as likely as all others to die of SIDS.
- SIDS is more common in lower socioeconomic groups.[65]

> IT IS ADVISABLE THAT AN INFANT BE PLACED SO IT SLEEPS ON ITS BACK, NOT ITS SIDE OR STOMACH. THIS WILL MAKE IT LESS LIKELY THAT BREATHING WILL BE OBSTRUCTED.

SLEEP DISORDERS ASSOCIATED WITH MEDICAL OR PSYCHIATRIC DISORDERS

There are a large number of medical and psychiatric disorders that produce problems with sleep and wakefulness. Among these are psychoses (such as schizophrenia), mood disorders (most often depression), anxiety and panic disorders, and alcoholism.

Forty-seven percent of Americans who experience severe insomnia report a high level of emotional distress. In depressed people, abnormal sleep patterns are often associated with an overwhelming feeling of sadness, hopelessness, or guilt. Some depressed people awaken early and cannot sleep; others sleep too much to avoid daytime stress. Many depressed individuals complain of insomnia without recognizing they are depressed. If you have lost interest in work, family, or recreation, or if you have feelings of hopelessness, or harbor suicidal thoughts, you may be suffering from depression. Though your complaint may be insomnia, the underlying depression, not the insomnia, must be treated.[66]

If you are being treated for a psychosis, a mood, anxiety, or panic disorder, or alcoholism, it is suggested that you discuss any sleep problems or daytime sleepiness with your primary-care physician. Often medications prescribed for treatment of these disorders will affect your sleep-wake cycle and alertness. Behavior therapies and new pharmacotherapies show promise in treating these problems, while at the same time not interfering with the need to get adequate sleep nor with daytime alertness.

WHEN TO CALL THE SLEEP DOCTOR

It's important to remember that sleep disorders are not rare and that they can be very serious. The chart in Figure 13.1 can help you decide if it's time to seek professional help. If after reading this chapter you think that you or someone you know may be suffering from a sleep disorder, contact a physician who has been trained in sleep medicine, or consult the list of sleep disorders clinics in Appendix D to find a source of professional help.

When to Seek Help

If you're routinely robbed of a good night's rest, you may have a sleep disorder. This chart lists symptoms associated with several common sleep problems. For each symptom you have, decide how severely or how frequently it affects you, on a 10-point scale. Then check the chart to see whether you should seek treatment. If you experience two or more symptoms, consider moving up to the next recommendation level.

If your family doctor's suggested remedies don't improve your sleep after a reasonable period, or if your main problem is daytime sleepiness, ask for a referral to a sleep disorders center for an evaluation.

For a list of accredited sleep centers, write to the American Sleep Disorders Association at 1610 14th St. NW, Suite 300, Rochester, MN 55901, or check its website (http://www.wisc.edu/asda).

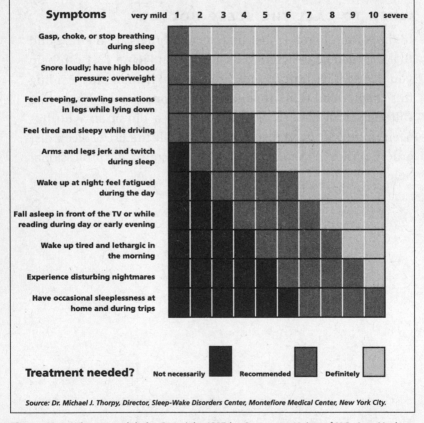

Source: Dr. Michael J. Thorpy, Director, Sleep-Wake Disorders Center, Montefiore Medical Center, New York City.

Figure 13.1. When to seek help. Copyright 1997 by Consumers Union of U.S., Inc., Yonkers, NY 10703-1057. Reprinted by permission from *Consumer Reports,* March 1997.

And if tonight my soul
may find her peace in sleep,
and sink in good oblivion,
and in the morning wake
like a new-opened flower
then I have been dipped again in God,
and new-created.

—D. H. LAWRENCE, "Shadows"

APPENDIXES

PEAK PERFORMANCE SLEEP LOGS

Peak Performance Sleep Log   Week 1   Name: _____

Copyright © 1998, Dr. James B. Maas

Every morning at breakfast fill out the chart for the previous day and night.
For example, on Monday morning you should complete the "Sunday" column.

Nights:	Sunday	Monday	Tuesday	Wednesday	Thursday	Friday	Saturday
What time did you turn your lights out?							
What time did you get up this morning?							
How many total hours did you sleep?							
How many times did you wake up during the night?							
Rate the quality of your sleep last night. 1 = terrible to 5 = great							
Did you avoid taking a nap yesterday?	Yes ☐ No ☐	Yes ☐ No ☐	Yes ☐ No ☐	Yes ☐ No ☐	Yes ☐ No ☐	Yes ☐ No ☐	Yes ☐ No ☐
Did you avoid caffeine after 6 P.M.?	Yes ☐ No ☐	Yes ☐ No ☐	Yes ☐ No ☐	Yes ☐ No ☐	Yes ☐ No ☐	Yes ☐ No ☐	Yes ☐ No ☐
Did you avoid alcohol after 6 P.M.?	Yes ☐ No ☐	Yes ☐ No ☐	Yes ☐ No ☐	Yes ☐ No ☐	Yes ☐ No ☐	Yes ☐ No ☐	Yes ☐ No ☐
Did you do anything to reduce stress yesterday?	Yes ☐ No ☐	Yes ☐ No ☐	Yes ☐ No ☐	Yes ☐ No ☐	Yes ☐ No ☐	Yes ☐ No ☐	Yes ☐ No ☐
Did you avoid sleeping medications?	Yes ☐ No ☐	Yes ☐ No ☐	Yes ☐ No ☐	Yes ☐ No ☐	Yes ☐ No ☐	Yes ☐ No ☐	Yes ☐ No ☐
Was your bedroom quiet, dark, and cool?	Yes ☐ No ☐	Yes ☐ No ☐	Yes ☐ No ☐	Yes ☐ No ☐	Yes ☐ No ☐	Yes ☐ No ☐	Yes ☐ No ☐
Did you do anything to relax before falling asleep?	Yes ☐ No ☐	Yes ☐ No ☐	Yes ☐ No ☐	Yes ☐ No ☐	Yes ☐ No ☐	Yes ☐ No ☐	Yes ☐ No ☐
Did you eat a balanced diet yesterday?	Yes ☐ No ☐	Yes ☐ No ☐	Yes ☐ No ☐	Yes ☐ No ☐	Yes ☐ No ☐	Yes ☐ No ☐	Yes ☐ No ☐
Did you exercise yesterday?	Yes ☐ No ☐	Yes ☐ No ☐	Yes ☐ No ☐	Yes ☐ No ☐	Yes ☐ No ☐	Yes ☐ No ☐	Yes ☐ No ☐
How alert and energetic did you feel during the day? 1 = sleepy, tired to 5 = fully alert, energetic							

How are you doing? To be prepared for peak performance (5's in the last row):
1. You should be getting close to eight hours of sleep each night.
2. Your sleep and wake times should not change between weekdays and weekends.
3. Your sleep should be continuous, not fragmented.
4. Your sleep should be restful.
5. The answers to all the yes-or-no questions should be yes.

Peak Performance Sleep Log   Week 2   Name: _____

Copyright © 1998, Dr. James B. Maas

Every morning at breakfast fill out the chart for the previous day and night.
For example, on Monday morning you should complete the "Sunday" column.

Nights:	Sunday	Monday	Tuesday	Wednesday	Thursday	Friday	Saturday
What time did you turn your lights out?							
What time did you get up this morning?							
How many total hours did you sleep?							
How many times did you wake up during the night?							
Rate the quality of your sleep last night. 1 = terrible to 5 = great							
Did you avoid taking a nap yesterday?	Yes ☐ No ☐	Yes ☐ No ☐	Yes ☐ No ☐	Yes ☐ No ☐	Yes ☐ No ☐	Yes ☐ No ☐	Yes ☐ No ☐
Did you avoid caffeine after 6 P.M.?	Yes ☐ No ☐	Yes ☐ No ☐	Yes ☐ No ☐	Yes ☐ No ☐	Yes ☐ No ☐	Yes ☐ No ☐	Yes ☐ No ☐
Did you avoid alcohol after 6 P.M.?	Yes ☐ No ☐	Yes ☐ No ☐	Yes ☐ No ☐	Yes ☐ No ☐	Yes ☐ No ☐	Yes ☐ No ☐	Yes ☐ No ☐
Did you do anything to reduce stress yesterday?	Yes ☐ No ☐	Yes ☐ No ☐	Yes ☐ No ☐	Yes ☐ No ☐	Yes ☐ No ☐	Yes ☐ No ☐	Yes ☐ No ☐
Did you avoid sleeping medications?	Yes ☐ No ☐	Yes ☐ No ☐	Yes ☐ No ☐	Yes ☐ No ☐	Yes ☐ No ☐	Yes ☐ No ☐	Yes ☐ No ☐
Was your bedroom quiet, dark, and cool?	Yes ☐ No ☐	Yes ☐ No ☐	Yes ☐ No ☐	Yes ☐ No ☐	Yes ☐ No ☐	Yes ☐ No ☐	Yes ☐ No ☐
Did you do anything to relax before falling asleep?	Yes ☐ No ☐	Yes ☐ No ☐	Yes ☐ No ☐	Yes ☐ No ☐	Yes ☐ No ☐	Yes ☐ No ☐	Yes ☐ No ☐
Did you eat a balanced diet yesterday?	Yes ☐ No ☐	Yes ☐ No ☐	Yes ☐ No ☐	Yes ☐ No ☐	Yes ☐ No ☐	Yes ☐ No ☐	Yes ☐ No ☐
Did you exercise yesterday?	Yes ☐ No ☐	Yes ☐ No ☐	Yes ☐ No ☐	Yes ☐ No ☐	Yes ☐ No ☐	Yes ☐ No ☐	Yes ☐ No ☐
How alert and energetic did you feel during the day? 1 = sleepy, tired to 5 = fully alert, energetic							

How are you doing? To be prepared for peak performance (5's in the last row):

1. You should be getting close to eight hours of sleep each night.
2. Your sleep and wake times should not change between weekdays and weekends.
3. Your sleep should be continuous, not fragmented.
4. Your sleep should be restful.
5. The answers to all the yes-or-no questions should be yes.

Peak Performance Sleep Log Week 3 Name: _____

Copyright © 1998, Dr. James B. Maas

Every morning at breakfast fill out the chart for the previous day and night.
For example, on Monday morning you should complete the "Sunday" column.

Nights:	Sunday	Monday	Tuesday	Wednesday	Thursday	Friday	Saturday
What time did you turn your lights out?							
What time did you get up this morning?							
How many total hours did you sleep?							
How many times did you wake up during the night?							
Rate the quality of your sleep last night. 1 = terrible to 5 = great							
Did you avoid taking a nap yesterday?	Yes ☐ No ☐	Yes ☐ No ☐	Yes ☐ No ☐	Yes ☐ No ☐	Yes ☐ No ☐	Yes ☐ No ☐	Yes ☐ No ☐
Did you avoid caffeine after 6 P.M.?	Yes ☐ No ☐	Yes ☐ No ☐	Yes ☐ No ☐	Yes ☐ No ☐	Yes ☐ No ☐	Yes ☐ No ☐	Yes ☐ No ☐
Did you avoid alcohol after 6 P.M.?	Yes ☐ No ☐	Yes ☐ No ☐	Yes ☐ No ☐	Yes ☐ No ☐	Yes ☐ No ☐	Yes ☐ No ☐	Yes ☐ No ☐
Did you do anything to reduce stress yesterday?	Yes ☐ No ☐	Yes ☐ No ☐	Yes ☐ No ☐	Yes ☐ No ☐	Yes ☐ No ☐	Yes ☐ No ☐	Yes ☐ No ☐
Did you avoid sleeping medications?	Yes ☐ No ☐	Yes ☐ No ☐	Yes ☐ No ☐	Yes ☐ No ☐	Yes ☐ No ☐	Yes ☐ No ☐	Yes ☐ No ☐
Was your bedroom quiet, dark, and cool?	Yes ☐ No ☐	Yes ☐ No ☐	Yes ☐ No ☐	Yes ☐ No ☐	Yes ☐ No ☐	Yes ☐ No ☐	Yes ☐ No ☐
Did you do anything to relax before falling asleep?	Yes ☐ No ☐	Yes ☐ No ☐	Yes ☐ No ☐	Yes ☐ No ☐	Yes ☐ No ☐	Yes ☐ No ☐	Yes ☐ No ☐
Did you eat a balanced diet yesterday?	Yes ☐ No ☐	Yes ☐ No ☐	Yes ☐ No ☐	Yes ☐ No ☐	Yes ☐ No ☐	Yes ☐ No ☐	Yes ☐ No ☐
Did you exercise yesterday?	Yes ☐ No ☐	Yes ☐ No ☐	Yes ☐ No ☐	Yes ☐ No ☐	Yes ☐ No ☐	Yes ☐ No ☐	Yes ☐ No ☐
How alert and energetic did you feel during the day? 1 = sleepy, tired to 5 = fully alert, energetic							

How are you doing? To be prepared for peak performance (5's in the last row):
1. You should be getting close to eight hours of sleep each night.
2. Your sleep and wake times should not change between weekdays and weekends.
3. Your sleep should be continuous, not fragmented.
4. Your sleep should be restful.
5. The answers to all the yes-or-no questions should be yes.

Peak Performance Sleep Log   Week 4   Name: _____

Copyright © 1998, Dr. James B. Maas

Every morning at breakfast fill out the chart for the previous day and night.
For example, on Monday morning you should complete the "Sunday" column.

Nights:	Sunday	Monday	Tuesday	Wednesday	Thursday	Friday	Saturday
What time did you turn your lights out?							
What time did you get up this morning?							
How many total hours did you sleep?							
How many times did you wake up during the night?							
Rate the quality of your sleep last night. 1 = terrible to 5 = great							
Did you avoid taking a nap yesterday?	Yes ☐ No ☐	Yes ☐ No ☐	Yes ☐ No ☐	Yes ☐ No ☐	Yes ☐ No ☐	Yes ☐ No ☐	Yes ☐ No ☐
Did you avoid caffeine after 6 P.M.?	Yes ☐ No ☐	Yes ☐ No ☐	Yes ☐ No ☐	Yes ☐ No ☐	Yes ☐ No ☐	Yes ☐ No ☐	Yes ☐ No ☐
Did you avoid alcohol after 6 P.M.?	Yes ☐ No ☐	Yes ☐ No ☐	Yes ☐ No ☐	Yes ☐ No ☐	Yes ☐ No ☐	Yes ☐ No ☐	Yes ☐ No ☐
Did you do anything to reduce stress yesterday?	Yes ☐ No ☐	Yes ☐ No ☐	Yes ☐ No ☐	Yes ☐ No ☐	Yes ☐ No ☐	Yes ☐ No ☐	Yes ☐ No ☐
Did you avoid sleeping medications?	Yes ☐ No ☐	Yes ☐ No ☐	Yes ☐ No ☐	Yes ☐ No ☐	Yes ☐ No ☐	Yes ☐ No ☐	Yes ☐ No ☐
Was your bedroom quiet, dark, and cool?	Yes ☐ No ☐	Yes ☐ No ☐	Yes ☐ No ☐	Yes ☐ No ☐	Yes ☐ No ☐	Yes ☐ No ☐	Yes ☐ No ☐
Did you do anything to relax before falling asleep?	Yes ☐ No ☐	Yes ☐ No ☐	Yes ☐ No ☐	Yes ☐ No ☐	Yes ☐ No ☐	Yes ☐ No ☐	Yes ☐ No ☐
Did you eat a balanced diet yesterday?	Yes ☐ No ☐	Yes ☐ No ☐	Yes ☐ No ☐	Yes ☐ No ☐	Yes ☐ No ☐	Yes ☐ No ☐	Yes ☐ No ☐
Did you exercise yesterday?	Yes ☐ No ☐	Yes ☐ No ☐	Yes ☐ No ☐	Yes ☐ No ☐	Yes ☐ No ☐	Yes ☐ No ☐	Yes ☐ No ☐
How alert and energetic did you feel during the day? 1 = sleepy, tired to 5 = fully alert, energetic							

How are you doing? To be prepared for peak performance (5's in the last row):
1. You should be getting close to eight hours of sleep each night.
2. Your sleep and wake times should not change between weekdays and weekends.
3. Your sleep should be continuous, not fragmented.
4. Your sleep should be restful.
5. The answers to all the yes-or-no questions should be yes.

Peak Performance Sleep Log　　　Week 5　　　Name: _____

Copyright © 1998, Dr. James B. Maas

Every morning at breakfast fill out the chart for the previous day and night.
For example, on Monday morning you should complete the "Sunday" column.

Nights:	Sunday	Monday	Tuesday	Wednesday	Thursday	Friday	Saturday
What time did you turn your lights out?							
What time did you get up this morning?							
How many total hours did you sleep?							
How many times did you wake up during the night?							
Rate the quality of your sleep last night. 1 = terrible to 5 = great							
Did you avoid taking a nap yesterday?	Yes ☐ No ☐	Yes ☐ No ☐	Yes ☐ No ☐	Yes ☐ No ☐	Yes ☐ No ☐	Yes ☐ No ☐	Yes ☐ No ☐
Did you avoid caffeine after 6 P.M.?	Yes ☐ No ☐	Yes ☐ No ☐	Yes ☐ No ☐	Yes ☐ No ☐	Yes ☐ No ☐	Yes ☐ No ☐	Yes ☐ No ☐
Did you avoid alcohol after 6 P.M.?	Yes ☐ No ☐	Yes ☐ No ☐	Yes ☐ No ☐	Yes ☐ No ☐	Yes ☐ No ☐	Yes ☐ No ☐	Yes ☐ No ☐
Did you do anything to reduce stress yesterday?	Yes ☐ No ☐	Yes ☐ No ☐	Yes ☐ No ☐	Yes ☐ No ☐	Yes ☐ No ☐	Yes ☐ No ☐	Yes ☐ No ☐
Did you avoid sleeping medications?	Yes ☐ No ☐	Yes ☐ No ☐	Yes ☐ No ☐	Yes ☐ No ☐	Yes ☐ No ☐	Yes ☐ No ☐	Yes ☐ No ☐
Was your bedroom quiet, dark, and cool?	Yes ☐ No ☐	Yes ☐ No ☐	Yes ☐ No ☐	Yes ☐ No ☐	Yes ☐ No ☐	Yes ☐ No ☐	Yes ☐ No ☐
Did you do anything to relax before falling asleep?	Yes ☐ No ☐	Yes ☐ No ☐	Yes ☐ No ☐	Yes ☐ No ☐	Yes ☐ No ☐	Yes ☐ No ☐	Yes ☐ No ☐
Did you eat a balanced diet yesterday?	Yes ☐ No ☐	Yes ☐ No ☐	Yes ☐ No ☐	Yes ☐ No ☐	Yes ☐ No ☐	Yes ☐ No ☐	Yes ☐ No ☐
Did you exercise yesterday?	Yes ☐ No ☐	Yes ☐ No ☐	Yes ☐ No ☐	Yes ☐ No ☐	Yes ☐ No ☐	Yes ☐ No ☐	Yes ☐ No ☐
How alert and energetic did you feel during the day? 1 = sleepy, tired to 5 = fully alert, energetic							

How are you doing? To be prepared for peak performance (5's in the last row):
1. You should be getting close to eight hours of sleep each night.
2. Your sleep and wake times should not change between weekdays and weekends.
3. Your sleep should be continuous, not fragmented.
4. Your sleep should be restful.
5. The answers to all the yes-or-no questions should be yes.

Peak Performance Sleep Log   Week 6   Name: _____

Copyright © 1998, Dr. James B. Maas

Every morning at breakfast fill out the chart for the previous day and night.
For example, on Monday morning you should complete the "Sunday" column.

Nights:	Sunday	Monday	Tuesday	Wednesday	Thursday	Friday	Saturday
What time did you turn your lights out?							
What time did you get up this morning?							
How many total hours did you sleep?							
How many times did you wake up during the night?							
Rate the quality of your sleep last night. 1 = terrible to 5 = great							
Did you avoid taking a nap yesterday?	Yes □ No □	Yes □ No □	Yes □ No □	Yes □ No □	Yes □ No □	Yes □ No □	Yes □ No □
Did you avoid caffeine after 6 P.M.?	Yes □ No □	Yes □ No □	Yes □ No □	Yes □ No □	Yes □ No □	Yes □ No □	Yes □ No □
Did you avoid alcohol after 6 P.M.?	Yes □ No □	Yes □ No □	Yes □ No □	Yes □ No □	Yes □ No □	Yes □ No □	Yes □ No □
Did you do anything to reduce stress yesterday?	Yes □ No □	Yes □ No □	Yes □ No □	Yes □ No □	Yes □ No □	Yes □ No □	Yes □ No □
Did you avoid sleeping medications?	Yes □ No □	Yes □ No □	Yes □ No □	Yes □ No □	Yes □ No □	Yes □ No □	Yes □ No □
Was your bedroom quiet, dark, and cool?	Yes □ No □	Yes □ No □	Yes □ No □	Yes □ No □	Yes □ No □	Yes □ No □	Yes □ No □
Did you do anything to relax before falling asleep?	Yes □ No □	Yes □ No □	Yes □ No □	Yes □ No □	Yes □ No □	Yes □ No □	Yes □ No □
Did you eat a balanced diet yesterday?	Yes □ No □	Yes □ No □	Yes □ No □	Yes □ No □	Yes □ No □	Yes □ No □	Yes □ No □
Did you exercise yesterday?	Yes □ No □	Yes □ No □	Yes □ No □	Yes □ No □	Yes □ No □	Yes □ No □	Yes □ No □
How alert and energetic did you feel during the day? 1 = sleepy, tired to 5 = fully alert, energetic							

How are you doing? To be prepared for peak performance (5's in the last row):
1. You should be getting close to eight hours of sleep each night.
2. Your sleep and wake times should not change between weekdays and weekends.
3. Your sleep should be continuous, not fragmented.
4. Your sleep should be restful.
5. The answers to all the yes-or-no questions should be yes.

SUGGESTED READINGS AND VIDEOTAPES ON SLEEP

PART I: SLEEP MATTERS

Caldwell, Paul. *Sleep*. Toronto, Canada: Key Porter Books Limited, 1995. General information on sleep featuring side bars, case studies, and diagrams.

Dement, William C. *The Sleepwatchers*. Palo Alto, Calif.: Stanford Alumni Association, 1992. An excellent resource on recent findings in sleep disorders medicine, sleep deprivation, and the purpose of sleep and dreams.

Mitler, Elizabeth A., and Merrill M. Mitler. *101 Questions About Sleep and Dreams*. Del Mar, Calif.: Wakefulness-Sleep Education and Research Foundation, 1996. Provides short, accurate answers to some of the most commonly asked questions.

PART II: THE POWER OF SLEEP

Carskadon, Mary A. *The Encyclopedia of Sleep and Dreaming*. New York: Macmillan, 1993. Covers in detail absolutely everything you have ever wanted to know about sleep or dreams.

Hobson, Allan J. *The Dreaming Brain*. New York: Basic Books, 1988. A research-based reference book on the history of dreams and dream interpretation.

———. *Sleep*. New York: Freeman, 1989. Explains the science of sleep and dreams for the layperson, providing a coherent view of modern sleep research. Uncovers the cellular and molecular mechanisms by which the brain regulates sleeping behavior.

PART III: PREPARING YOUR MIND FOR PEAK PERFORMANCE

You are currently reading the only book devoted specifically to this topic!

PART IV: COPING WITH SLEEP DEPRIVATION

Chapter 9: The Nod to Midday Naps

Coren, Stanley. *Sleep Thieves: An Eye-opening Exploration into the Science and Mysteries of Sleep.* New York: Free Press, 1996. Chronicles the need in our society to place more importance on the role of sleep.

Dinges, David, and Roger Broughton. *Sleep and Alertness: Chronological, Behavioral and Medical Aspects of Napping.* New York: Raven Press, 1990. The bible of research on napping.

Chapter 10: Surviving as a Shift Worker

Coleman, Richard M. *Wide Awake at 3:00 A.M.* New York: W. H. Freeman, 1986. A good brief reference on shift-work studies and jet lag.

Dotto, Lydia. *Losing Sleep.* New York: William Morrow, 1990. The best overview of sleep studies related to jobs and performance.

Moore-Ede, Martin, M.D. *The Twenty-four-Hour Society.* New York: Addison Wesley, 1993. Understanding human limits in a world that never stops.

Chapter 11: Reducing Travel Fatigue

Ehret, Charles F., and Lynne Waller Scanlon. *Overcoming Jet Lag.* New York: Berkley, 1983. Three-step comprehensive guide to combatting jet lag.

Mayes, Kathleen. *Beat Jet Lag.* London: Thorsons, 1991. Easy-to-read book on how to arrive alert and stay alert.

Oren, Dan A., et al. *How to Beat Jet Lag.* New York: Henry Holt, 1993. Step-by-step instructions for conquering jet lag and handling red-eye flights.

Chapter 12: Avoiding Family Sleep Traps

Cuthbertson, Joan, and Susie Schevill. *Helping Your Child Sleep Through the Night.* New York: Doubleday, 1985. Cites some interesting solutions to some of the problems (bed-wetting, sleepwalking, etc.) that often disrupt children's sleep.

Ferber, Richard. *Solve Your Child's Sleep Problems*. New York: Simon & Schuster, 1985. For parents whose children are experiencing sleep problems.

Ferber, Richard, and Meir Kryger. *Principles and Practices of Sleep Medicine in the Child*. New York: Saunders, 1995. A medical textbook that deals with troubled sleep of children.

Graber, Richard, with Paul Gouin. *How to Get a Good Night's Sleep*. Minneapolis: Chronimed Publishing, 1995. More than 100 ways to improve your sleep.

Hartmann, Ernest. *The Sleep Book: Understanding and Preventing Sleep Problems in People Over 50*. Washington, D.C.: American Association of Retired Persons, 1987.

Inlander, Charles B., and Cynthia K. Moran. *67 Ways to Good Sleep*. New York: Ballantine Books, 1995. Includes a broad range of topics ranging from proper diet and sleep environment to sleeping pills and disorders.

Kavey, Neil B. *50 Ways to Sleep Better*. Lincolnwood, Ill.: Publications International Ltd., 1996. Handy hints for restful nights. Published in association with the Sleep Disorders Center, Columbia-Presbyterian Medical Center, New York, N.Y.

Morgan, Kevin. *Sleep and Aging: A Research-Based Guide to Sleep in Later Life*. Baltimore, Md.: Johns Hopkins University Press, 1987.

PART V: WHEN TO CALL THE SLEEP DOCTOR

Ford, Norman. *The Sleep R_x*. Englewood Cliffs, N.J.: Prentice Hall, 1994. An excellent book on the architecture of sleep, self-diagnosis of insomnia problems, and suggested cures.

Fritz, Roger. *Sleep Disorders: America's Hidden Nightmare*. Naperville, Ill.: National Sleep Alert, Inc., 1993. A review of sleep deprivation and sleep disorders.

Hauri, Peter, and Shirley Linde. *No More Sleepless Nights*. New York: John Wiley, 1990. An excellent guide for adults with sleep problems.

Kryger, Meir H., Thomas Roth, and William C. Dement. *Principles and Practices of Sleep Medicine*. New York: Saunders, 1994. A medical textbook outlining methods used in the various treatments for sleep disorders.

Lavie, Peretz. *The Enchanted World of Sleep*. New Haven, Conn.: Yale University Press, 1996. A very readable survey of sleep research and sleep medicine.

Pascualy, Ralph A., and Sally Warren Soest. *Snoring and Sleep Apnea: Personal and Family Guide to Diagnosis and Treatment*. New York: Raven Press, 1994. Details on sleep apnea, as well as some handy hints about stopping snorers. Available either through the publisher or at medical bookstores. Order direct from Demos Vermande, 386 Park Avenue South, Suite 201, New York, NY 10016. Tel: 212-683-0072, or 800-532-8663.

Podell, Richard N. *Doctor, Why Am I So Tired?* New York: Pharos Books, 1987. Discusses plausible causes of incessant fatigue.

Regestein, Quentin R., et al., eds. *Sleep: Problems and Solutions.* Mount Vernon, N.Y.: Consumer Union Report Books, 1990.

Shapiro, Colin M. *Conquering Insomnia: An Illustrated Guide to Understanding Sleep and a Manual for Overcoming Sleep Disruption.* Hamilton, Ont.: Empowering Press, 1994.

Sweeney, Donald R. *Overcoming Insomnia.* New York: G. P. Putnam's Sons, 1989. After you've read it, we hope this book can put you to sleep!

Utley, Margaret Jones. *Narcolepsy: A Funny Disorder That's No Laughing Matter.* A great source of information on narcolepsy that is easy to read and includes the author's own experiences with the disease. To order, contact M. J. Utley, P.O. Box 1923, DeSoto, TX 75123-1923.

For further information about sleep and sleep disorders, contact the National Sleep Foundation, 1367 Connecticut Avenue NW, Suite 200, Washington, DC 20036, and visit the websites listed in Appendix C.

VIDEOTAPES ON SLEEP

Maas, James B. "Sleep Alert." Ithaca, N.Y.: Cornell University Psychology Film Unit, 1993. A nationally broadcast PBS special on the problem of sleep deprivation in America.

————. "Asleep in the Fast Lane: Our 24-Hour Society." Ithaca, N.Y.: Cornell University Psychology Film Unit, 1998. A documentary on how lack of sleep causes accidents and poor performance in our fast-paced society.

Maas, James B., and David H. Gluck. "Keep Us Awake." Ithaca, N.Y.: Cornell University Psychology Film Unit, 1978. A documentary film on the symptoms, causes, and treatment of narcolepsy.

————. "When Nights Are Longest." Ithaca, N.Y.: Cornell University Psychology Film Unit, 1983. A documentary on the symptoms, causes, and cures for the various syndromes that constitute the sleep disorder of insomnia.

All of the above videotapes, in VHS format, are available for purchase. Contact Dr. James B. Maas, 210 Uris Hall, Department of Psychology, Cornell University, Ithaca, NY 14853. Tel: 607-255-6266.

SLEEP ON THE INTERNET

THE SLEEP WELL
http://www-leland.stanford.edu/~dement/

A tremendous source of up-to-date information on all aspects of sleep. Maintained by Stanford University, a major center for sleep research. Includes a calendar of sleep-related events, an extensive bibliography, and the addresses of sleep disorders associations and help groups.

THE SLEEPNET HOME PAGE
http://www.sleepnet.com/

Everything you want to know about sleep but are too tired to ask. Provides some startling statistics as well as more than 130 rated hot links to other sleep websites.

NATIONAL SLEEP FOUNDATION
http://www.sleepfoundation.org/

This is a direct link to the best-known nonprofit foundation for sleep. Here you can get the answers to questions concerning sleeping pills or other medications.

ACCESS TO SLEEP AND PSYCHOLOGY
http://www.yahoo.com/Health/Medicine/Sleep_Medicine/

An academic science site: basic how-to and self-help regarding sleep research and psychology. Includes abstracts on melatonin and sleep, health, and development.

A GOOD NIGHT'S SLEEP
http://www.pacificcoast.com/pcf/healthy/gdslp.html

An informative summary on how much sleep you need, the basic architecture of a night's sleep, common sleep problems, naps, stress and sleep, and sleep and the life cycle. Ten ways to help you sleep for a better tomorrow. Discusses exercise, nutrition, the bedroom environment, and sleep schedules. Advice about mattresses, pillows, and being comfortable sharing a bed.

HOW TO FEEL RESTED ON TOO LITTLE SLEEP
http://www.homearts.com/rb/health/04sleef1.html

Discusses the nationwide dilemma of sleep deprivation, the effects of sleep loss, and ways to combat the consequences.

CHILDREN'S SLEEP PROBLEMS
http://www.psych.med.umich.edu/web/aacap/factsFam/sleep.html

Is your child's sleep disrupted by bed-wetting, night terrors, or sleepwalking? Some information and suggestions are provided.

BRAIN INFORMATION SERVICES
http://bisleep.medsch.ucla.edu/

An excellent resource for those who are involved in the research and/or treatment of sleep disorders. Provides links to sleep support organizations, the World Federation of Sleep Research Societies, sleep labs, newsletters, articles, and more.

THE SLEEP MEDICINE HOME PAGE
http://www.cloud9.net/~thorpy/

This website provides links to information sites concerning virtually every sleep disorder, and to listings of sleep physicians, sleep disorders centers, research centers, sleep-related foundations, and professional associations.

THE SCHOOL OF SLEEP MEDICINE
http://www.sleepnet.com/ssm.html

The School of Sleep Medicine (SSM) was founded to provide high-quality education for the medical community worldwide. Its purpose is to increase awareness among physicians and other health-care professionals of the nature and treatment of sleep disorders

NARCOLEPSY
http://www.hia.com/hia/narcoctr/

The Stanford University Center for Narcolepsy home page. Interests of the research center range from finding better pharmacological treatments for human patients to isolating the genes for narcolepsy.

SLEEP APNEA
http://medinfo-online.com/sleep.html

Ten brief tips to a better night's sleep. Also, an in-depth look at the symptoms, diagnosis, and treatment of obstructive sleep apnea (OSA).

NAPTIME DREAMS
http://www.bluemarble.net/~amyloo/dream.html

A report on the duration, content, rate of recall, and enjoyment of dreams during naps in contrast to dreams experienced during the night.

SLEEP AND DREAMS
http://www.yahoo.com/science/psychology/sleep_and_dreams/

A dream site providing links to information on dream analysis, lucid dreams, and the Association for the Study of Dreams.

PHANTOM SLEEP PAGE
http://www.newtechpub.com/phantom/

A site for those who think they may be suffering from sleep apnea. Includes information on snoring and other sleep problems as well as a free self-scoring sleep apnea quiz.

SLEEP, DREAMS AND WAKEFULNESS—UNIVERSITY OF LYON
http://ura1195-6.univ-lyon1.fr/home.html

A collection of complete (primarily academic) articles related to sleep. Includes a database of 12,000 references.

SHUTEYE ONLINE
http://www.shuteye.com/index.html

An interactive website launched by Searle helps visitors learn more about sleep problems, track their sleep habits, and pursue effective treatment and preventive measures.

SLEEP DISORDERS CENTERS

American Sleep Disorders Association-accredited member centers and laboratories

An asterisk (*) denotes an accredited sleep disorders laboratory for sleep-related breathing disorders; all other programs are accredited full-service sleep disorders centers. A sleep disorders center is a medical facility providing clinical diagnostic services and treatment to
patients who present with symptoms or features that suggest the presence of a sleep disorder. A laboratory for sleep-related breathing disorders provides diagnostic and treatment services limited to sleep-related breathing disorders, such as obstructive sleep apnea syndrome.

ALABAMA

Sleep-Related Breathing Disorders Lab*
Athens-Limestone Hospital
700 West Market Street
PO Box 999
Athens, AL 35612
Andy Jackson
Cherri Walker
256-771-REST (7378)
Fax: 256-233-9575

Brookwood Sleep Disorders Center
Brookwood Medical Center
2010 Brookwood Medical Center
  Drive
Birmingham, AL 35209
Robert C. Doekel, M.D.
Regina McClain, RPSGT, Sleep
  Center Coordinator
205-877-2403 Fax: 205-877-1663

Sleep Disorders Lab*
Carraway Methodist Medical Center
1600 Carraway Boulevard
Birmingham, AL 35234
Kurvilla George
Aneshia Williams
205-502-6164 Fax: 205-502-5210

Sleep-Wake Disorders Center
University of Alabama at
Birmingham
1713 6th Avenue South
CPM Building, Room 270
Birmingham, AL 35233-0018
Susan Harding, M.D.
Vernon Pegram, Ph.D.
Len Shigley, RPSGT,
Technical Director
205-934-7110 Fax: 205-934-6870
Email: lshigley@uabmc.edu

Sleep Disorders Center of Alabama, Inc.
790 Montclair Road, Suite 200
Birmingham, AL 35213
Vernon Pegram, Ph.D.
Robert C. Doekel, M.D.
205-599-1020 Fax: 205-599-1029

Princeton Sleep/Wake Disorders Center
Baptist Medical Center Princeton
701 Princeton Avenue SW, Suite 50
PO Box II
Birmingham, AL 35211-1399
Cary Evans
205-783-7378 Fax: 205-783-7386

Breathing Related Sleep Disorders Center*
Marshall Medical Center South
601A Corley Avenue
PO Box 758
Boaz, AL 35957
Lori Johnson, RPSGT, Coordinator
Robert Doekel, Jr., M.D.

256-593-1226 Fax:256-593-9945
Email: lori.johnson@mmcs.org

Sleep Disorders Center
Cullman Regional Medical Center
1912 Alabama Highway 157
Cullman, AL 35056-1108
G. Scott Warner, M.D., Medical
Director
Lisa Nelson Barnett, RPSGT,
Administrative Director
256-737-2140 Fax: 256-737-2261

Decatur General Sleep Disorders Center
1201 7th Street Southeast
Decatur, AL 35601
Edward M. Turpin, M.D.
Marc A. Hays, RRT
256-340-2558 Fax: 256-340-2566

Sleep-Wake Disorders Center
Flowers Hospital
4370 West Main Street
PO Box 6907
Dothan, AL 36302
Ronald C. Kornegay, RPSGT,
Ann B. McDowell, M.D.
Alan Purvis, M.D.
David Davis, M.D.
334-793-5000 x1685
Fax: 334-615-7213

Thomas Hospital Sleep Services*
Thomas Hospital
188 Hospital Drive, Suite 201
Fairhope, AL 36532
William E. Goetter, M.D. FCCP
James J. Griffin, M.D.
334-990-1940 Fax: 334-990-1941

ECM Sleep Disorders Lab*
Eliza Coffee Memorial Hospital
205 Marengo Street
PO Box 818
Florence, AL 35631
Felix Morris, M.D.
Byron Jamerson, RPSGT
256-768-9153 Fax: 256-740-8524

Sleep Diagnostics of Northeast Alabama
for Breathing Related Disorders
Gadsden Regional Medical Center*
1007 Goodyear Avenue
Gadsden, AL 35903
Denise J. Barton, RRT, RPSGT
256-494-4551 Fax: 256-494-4602

The Sleep Center
Huntsville Hospital
911 Big Cove
Huntsville, AL 35801
Paul LeGrand, M.D.
Debra J. Vaughn, MBA, RRT, RPSGT
256-517-8553 Fax: 256-533-8388
Email: paull@md.hhsys.org

The Crestwood Center for Sleep Disorders
250 Chateau Drive Suite 235
Huntsville, AL 35801
Alan H. Arrington, M.D.
Thomas H. Arrington, RPSGT
205-880-4710 Fax: 205-880-4708

Southeast Regional Center for Sleep/Wake
Disorders Springhill Memorial Hospital
3719 Dauphin Street
Mobile, AL 36608
Lawrence S. Schoen, Ph.D.
334-460-5319 Fax: 334-460-5464

USA Knollwood Sleep Disorders Center
University of South Alabama-
Knollwood Park Hospital
5644 Girby Road
Mobile, AL 36693-3398
William A. Broughton, M.D.
334-660-5757 Fax: 334-660-5254
Email: jetbrou@ibm.net

Sleep Disorders Center
Mobile Infirmary Medical Center
PO Box 2144
Mobile, AL 36652
Robert Dawkins, Ph.D., M.P.H.
334-435-5559 Fax: 334-435-5222

Sleep Disorders Center
Baptist Medical Center
2105 East South Boulevard
Montgomery, AL 36116-2498
David P. Franco, M.D.
Tammy Taylor, RPSGT
334-286-3252 Fax: 334-286-3108

Sleep Disorders Lab*
East Alabama Medical Center
2000 Pepperell Parkway
Opelika, AL 36801-5452
Nancy Strickland, RRT
Steven E.Dekich, M.D.
334-705-2404 Fax: 334-705-2403
Email: nancy_strickland@eamc.org
Website: www.eamc.org

Sleep Disorders Lab*
Helen Keller Hospital
PO Box 610
Sheffield, AL 35660
Paul Schuler, M.D.
Ronda Hood, RRT
256-386-4191 Fax: 256-386-4323
Email: rhood@helenkeller.com

Tuscaloosa Clinic Sleep Center
701 University Boulevard
East Tuscaloosa, AL 35401
Richard M. Snow, M.D., FCCP
205-349-4043

ALASKA

Sleep Disorders Center
Providence Alaska Medical Center
3200 Providence Drive
PO Box 196604
Anchorage, AK 99519-6604
Anne H. Morris, M.D., Medical
   Director
Katie Boyle Colborn, RPSGT,
   Clinical Manager
907-261-3650 Fax: 907-261-4810
Email: anne-morris@sppha-
   03.ccmail.compuserve.com

ARIZONA

Samaritan Regional Sleep Disorders Program
Thunderbird Samaritan Medical
   Center
5555 West Thunderbird Road
Glendale, AZ 85306-4622
Bernard E. Levine, M.D.
Stephen Anthony, M.D.
Connie Boker, RPSGT
602-588-4800 Fax: 602-588-4810

Samaritan Regional Sleep Disorders Program
Desert Samaritan Medical Center
1400 South Dobson Road
Mesa, AZ 85202
Paul Barnard, M.D.
Tom Munzlinger, BS, RPSGT
602-835-3684 Fax: 602-835-8788

Samaritan Regional Sleep Disorders Program
Good Samaritan Regional Medical
   Center
1111 East McDowell Road
Phoenix, AZ 85006
Bernard E. Levine, M.D.
David Baratz, M.D.
Connie Boker, RPSGT
602-239-5815 Fax: 602-239-2129

Sleep Disorders Center
Scottsdale Healthcare
9003 East Shea Boulevard
Scottsdale, AZ 85260
Jeffrey S. Gitt, D.O.
Sharon E. Cichocki, RPSGT
602-860-3200 Fax: 602-860-3251

Sleep Disorders Center
University of Arizona
1501 North Campbell Avenue
Tucson, AZ 85724
Stuart F. Quan, M.D.
520-694-6112 or 520-626-6115
Fax: 520-694-2515
Email: squan@sneeze.resp-sci.
   arizona.edu

ARKANSAS

Sleep Disorders Center
Washington Regional Medical
   Center
1125 North College Avenue
Fayetteville, AR 72703
David L. Brown, M.D., Director
William A. Rivers, RPSGT,
   Coordinator
501-713-1272 Fax: 501-713-1190

Pediatric Sleep Disorders
Arkansas Children's Hospital
800 Marshall Street
Little Rock, AR 72202-3591
May Griebel, M.D.
Linda Rhodes, EMT, RPSGT
501-320-1893 Fax: 501-320-6878
Email: lkr@exchange.ach.uams.edu

Sleep Disorders Center
Baptist Medical Center
9601 I-630, Exit 7
Little Rock, AR 72205-7299
David Davila, M.D.
Buddy Marshall, CRTT, RPSGT
501-202-1902 Fax: 501-202-1874
Email: dgdavila@baptist-health.org

CALIFORNIA

**Western Medical Centers' Sleep Disorders
   Center**
1101 South Anaheim Boulevard
Anaheim, CA 92805
Clyde Dos Santos, M.D., Medical
   Director
Deborah Kerr, Director
714-491-1159 Fax: 714-563-2865

Sleep Center
Mercy San Juan Hospital
6401 Coyle Avenue, Suite 109
Carmichael, CA 95608
Janice K. Herrmann, RPSGT, MA
Richard Stack, M.D.
916-864-5874 Fax: 916-864-5870
Email: rstack@sma.com

Sleep Disorders Institute
St. Jude Medical Center
1915 Sunny Crest Drive
Fullerton, CA 92835
Louis J. McNabb, M.D.
Justine Petrie, M.D.
Robert Roethe, M.D.
714-446-7240 Fax: 714-446-7245

**Glendale Adventist Medical Center Sleep
   Disorders Center**
Glendale Adventist Medical Center
1509 Wilson Terrace
Glendale, CA 91206
David A. Thompson, M.D.
Kathy Cavander
818-409-8323 Fax: 818-546-5625

Pacific Sleep Medicine Services
La Jolla Center
9834 Genesee Avenue, Suite 328
La Jolla, CA 92037-1223
Milton Erman, M.D.
Stuart Menn, M.D.
619-657-0550 Fax: 619-657-0559
Email: merman@scripps.edu and
   75620.37@compuserve.com
Website: www.sleepmedservices.com

Sleep Disorders Center
Grossmont Hospital
PO Box 158
La Mesa, CA 91944-0158
Ellie Hoey, RPSGT
619-644-4488 Fax: 619-644-4021

Loma Linda Sleep Disorders Center
Loma Linda University Community
   Medical Center
25333 Barton Road
Loma Linda, CA 92354
Ralph Downey, III, Ph.D.
Joanne MacQuarrie, RRT, RPSGT
909-478-6344 Fax: 909-478-6343

Sleep Disorders Center
Long Beach Memorial Medical
   Center
2801 Atlantic Avenue
PO Box 1428
Long Beach, CA 90801-1428
Stephen E. Brown, M.D.
Monir Kashani, RRT, RPSGT,
   Technical Coordinator
562-933-0208 Fax: 562-933-0201

UCLA Sleep Disorders Center
24-221 CHS, Box 957069
Los Angeles, CA 90095-7069
Frisca Yan-Go, M.D.
Jerald Simmons, M.D.
310-206-8005 Fax: 310-206-3348

Los Gatos Clinical Monitoring Center, Inc.
Sleep Disorders Center
555 Knowles Drive, Suite 218
Los Gatos, CA 95032
Tom Pace, RPSGT, Clinical
   Coordinator
Laughton Miles, M.D., Ph.D.
Augustin de la Pena, Ph.D.
Roger Smith, D.O.
408-341-2080 Fax: 408-341-2088
Email: cmc@sleepscape.com
Website: www.sleepscape.com

Sleep Disorders Center
Hoag Memorial Hospital
   Presbyterian
One Hoag Drive
PO Box 6100
Newport Beach, CA 92658-6100
Catherine L. Rain, Coordinator
Paul A. Selecky, M.D.
949-760-2070 Fax: 949-574-6297
Email: crain@hoaghospital.org
Website: www.hoag.org

Sleep Evaluation Center
Northridge Hospital Medical Center
18300 Roscoe Boulevard
Northridge, CA 91328
Jeremy Cole, M.D.
David Brandes, M.D.
Dennis McGinty, Ph.D.
Ron Szymusiak, Ph.D.
818-885-5344

California Center for Sleep Disorders
3012 Summit Street, South
   Building, 5th Floor
Oakland, CA 94609
Jerrold Kram, M.D.
Glenn Roldan, BS, RPSGT
510-834-8333 Fax: 510-834-4728

St. Joseph Hospital Sleep Disorders Center
1310 West Stewart Drive, Suite 403
Orange, CA 92868
Sarah Mosko, Ph.D.
714-771-8950 Fax: 714-744-8541

Sleep Disorders Center
University of California, Irvine
101 City Drive, Route 23
Orange, CA 92868
Peter A. Fotinakes, M.D.
714-456-5105 Fax: 714-456-7822

Premier Diagnostics, Inc.
1851 Holser Walk, Suite 210
Oxnard, CA 93030
Jerry Harris, RCP, RRT
Rebecca Palmieri, RCP, RRT, RN, BS
George Yu, M.D.
805-485-2633 Fax: 805-485-6650

Sleep Disorders Center
Huntington Memorial Hospital
100 West California Boulevard
PO Box 7013
Pasadena, CA 91109-7013
Steven Lenik, RPSGT
Charles A. Anderson, M.D.
Richard A. Shubin, M.D.
626-397-3061 Fax: 626-397-3211
Email: sleeplab@ix.netcom.com

**Sleep Disorders Center Doctors Medical
  Center – Pinole**
2151 Appian Way
Pinole, CA 94564-2578
Darlene Connolly, RN
Frederick Nachtwey, M.D.
Richard Sankary, M.D.
510-741-2525 and 800-640-9440
Fax: 510-724-2189

Sleep Disorders Center
Pomona Valley Hospital Medical
  Center
1798 North Garey Avenue
Pomona, CA 91767
Dennis Nicholson, M.D.
Fares Elghazi, M.D.
Robert Jones, M.D., FCCP
909-865-9587 Fax: 909-865-9969

The Center for Sleep Apnea*
Redding Medical Center
2701 Eureka Way, Suite 1I
Redding, CA 96001
Everett Trevor, M.D.
Jean Amari-Melancon, RPSGT
530-242-6821 Fax: 916-245-4116

Sequoia Sleep Disorders Center
Sequoia Hospital
170 Alameda de las Pulgas
Redwood City, CA 94062-2799
J. Al Reichert, RPSGT
Bernhard Votteri, M.D., Medical
  Director
650-367-5137 Fax: 650-363-5304
Email: sleep@sleepscene.com
Website: www.sleepscene.com

Sutter Sleep Disorders Center
650 Howe Avenue, Suite 910
Sacramento, CA 95825
Sue Van Duyn, RPSGT, RCP, BA
Lydia Wytrzes, M.D.
916-646-3300 Fax: 916-646-4603
Email: vandus@sutterhealth.org

UCDMC Sleep Disorders Center
University of California Davis
  Medical Center
2315 Stockton Boulevard, Room
  5305
Sacramento, CA 95817
Masud Seyal, M.D., Ph.D.
William Bonekat, D.O.
916-734-0256 Fax: 916-452-2739

Inland Sleep Center
401 East Highland Avenue, Suite
  552
San Bernardino, CA 92404
Sunil Arora, M.D.
909-883-8058 Fax: 909-881-4607

San Diego Sleep Disorders Center
1842 Third Avenue
San Diego, CA 92101
Renata Shafor, M.D.
619-235-0248 Fax: 619-544-0588
Email: shafor@znet.com

Mercy Sleep Disorders Center
Scripps Mercy Hospital
4077 Fifth Avenue
San Diego, CA 92103-2180
Alex Mercandetti, M.D., FCCP,
   Medical Director
Cheryl L. Spinweber, Ph.D., Clinical
   Director
619-260-7378 Fax: 619-686-3990

UCSF/Stanford Sleep Disorders Center
University of California, San
   Francisco
1600 Divisadero Street
San Francisco, CA 94115
David M. Claman, M.D.
Kimberly A. Trotter, MA, RPSGT
415-885-7886 Fax: 415-885-3650

**Stanford Health Services Sleep Clinic in
   San Francisco**
2340 Clay Street, Suite 237
San Francisco, CA 94115-1932
Bruce T. Adornato,M.D.
Christopher R. Brown, M.D.
Rowena Korobkin, M.D.
Clete Kushida, M.D., Ph.D.
415-923-3336 Fax: 415-923-3584

**The Sleep Disorders Center of Santa
   Barbara**
2410 Fletcher Avenue, Suite 201
Santa Barbara, CA 93105
Andrew S. Binder, M.D.
Laurie Laatsch, RPSGT
805-898-8845 Fax: 805-898-8848

Stanford Sleep Disorders Clinic
Stanford University Medical Center
401 Quarry Road
Stanford, CA 94305
Jed Black, M.D.
650-723-6601 Fax: 650-725-8910

Southern California Sleep Apnea Center*
Lombard Medical Group
2230 Lynn Road
Thousand Oaks, CA 91360
Ronald A. Popper, M.D., FCCP
805-449-1096 Fax: 805-497-1782

**Torrance Memorial Medical Center Sleep
   Disorders Center**
3330 West Lomita Boulevard
Torrance, CA 90505
Lawrence W. Kneisley,M.D.
310-517-4617 Fax: 310-784-4869

Sleep Disorders Laboratory*
Kaweah Delta District Hospital
400 West Mineral King Avenue
Visalia, CA 93291
William R. Winn, M.D.
Gregory C. Warner, M.D.
Larry Kellett, BS, RCPT, Clinical
   Coordinator
209-625-7338 Fax: 209-635-4059
Email: lkellsleep@aol.com

West Valley Sleep Disorders Center
7320 Woodlake Avenue, Suite 140
West Hills, CA 91307
Gordon Dowds, M.D., Medical
   Director
Pamela Pierce, Director
818-715-0096 Fax: 818-716-1875
Email: gordon@dowds.com

Sleep Disorders Center
Woodland Memorial Hospital
1325 Cottonwood Street
Woodland, CA 95695
Richard A. Beyer, M.D.
Marie Kearney, Manager
530-668-2695 Fax: 530-662-9174

COLORADO

National Jewish/University of Colorado
Sleep Center
1400 Jackson Street, A200
Denver, CO 80206
Robert D. Ballard, M.D.
303-398-1523

Sleep Disorders Center
Presbyterian St. Luke's Medical
Center
1719 East 19th Avenue
Denver, CO 80218
John Ruddy, M.D.
303-839-6049 Fax: 303-869-1815

Sleep Center of Southern Colorado
Parkview Medical Center
400 West Sixteenth Street
Pueblo, CO 81003
James Pagel, M.D.
Ron Fossceco, RRT, RPSGT
719-584-4659 Fax: 719-584-4929
Email: ronf@parkviewmc.com

CONNECTICUT

Danbury Hospital Sleep Disorders Center
Danbury Hospital
24 Hospital Avenue
Danbury, CT 06810
Arthur Kotch, M.D.
Arthur Spielman, Ph.D.
203-731-8033 Fax: 203-731-8628
Email: kotcha@danhosp.org

Yale Center for Sleep Disorders
Yale University School of Medicine
333 Cedar Street
PO Box 208057
New Haven, CT 06520-8057
Vahid Mohsenin, M.D.
203-737-5556 Fax: 203-453-0630
Email: sleep.disorders@yale.edu

Gaylord-Wallingford Sleep Disorders
Laboratory*
Gaylord Hospital
Gaylord Farms Road
Wallingford, CT 06492
Thomas Whelan, RPSGT
Vahid Mohsenin, M.D.
203-284-2853 Fax: 203-284-2746

DELAWARE

Sleep Disorders Center
Christiana Care Health Systems
4755 Ogletown-Stanton Road
PO Box 6001
Newark, DE 19718
John B. Townsend, III, M.D.
Thomas C. Mueller, M.D.
Mary Rose Hancock
302-428-4600 Fax: 302-733-2533

Sleep Disorders Center
Christiana Care Health Services
Wilmington Hospital
501 West 14th Street
Wilmington, DE 19899
John B. Townsend, III, M.D.
Thomas C. Mueller, M.D.
302-428-4600 Fax: 302-733-2533
Email: hancock.m@christianacare.org

DISTRICT OF COLUMBIA

Sleep Disorders Center
5 Main Hospital, Georgetown
   University Hospital
3800 Reservoir Road NW
Washington, D.C. 20007-2197
Marilyn L. Faucette, RPSGT
Anne O'Donnell, M.D.
Kenneth Plotkin, M.D.
Richard E. Waldhorn, M.D., Medical
   Director
202-784-3610 Fax: 202-784-2920

**Sibley Memorial Hospital Sleep Disorders
   Center**
5255 Loughboro Road NW
Washington, D.C. 20016
David N. F. Fairbanks, M.D.
Samuel J. Potolicchio, M.D.
202-364-7676 Fax: 202-362-9378

FLORIDA

Boca Raton Sleep Disorders Center
899 Meadows Road, Suite 101
Boca Raton, FL 33486
Natalio J. Chediak, M.D.
Sheila R. Shafer, CMA
561-750-9881 Fax: 561-750-9644

Sleep Disorder Laboratory*
Broward General Medical Center
1600 South Andrews Avenue
Fort Lauderdale, FL 33316
Glenn R. Singer, M.D.
954-355-5534 Fax: 954-355-4848
Email: gsinger@nbhd.org
Website: www.nbhd.org/facility/
   sleep.htm

Mayo Sleep Disorders Center
Mayo Clinic Jacksonville
4500 San Pablo Road
Jacksonville, FL 32224
Paul Fredrickson, M.D.
Joseph Kaplan, M.D.
904-953-7287 Fax: 904-953-7388

Watson Clinic Sleep Disorders Center
The Watson Clinic, LLP
1600 Lakeland Hills Boulevard
PO Box 95000
Lakeland, FL 33804-5000
Eberto Pineiro, M.D.
941-680-7627 Fax: 941-680-7430

Atlantic Sleep Disorders Center
1401 South Apollo Boulevard,
   Suite A
Melbourne, FL 32901
Dennis K. King, M.D.
407-952-5191 Fax:407-952-7262

Sleep Disorders Center
Mt. Sinai Medical Center
4300 Alton Road
Miami Beach, FL 33140
Alejandro D. Chediak, M.D.
305-674-2613

Sleep Disorders Center
Miami Children's Hospital
6125 Southwest 31st Street
Miami, FL 33155
Marcel J. Deray, M.D.
305-669-7136 Fax: 305-669-6472
Email: maderay@aol.com

University of Miami School of Medicine
JMH and VA Medical Center Sleep
  Disorders Center, Department of
  Neurology (D4-5)
PO Box 016960
Miami, FL 33101
Bruce Nolan, M.D.
305-324-3371

**Munroe Regional Medical Center Sleep
  Laboratory***
Munroe Regional Medical Center
131 Southwest 15th Street
Ocala, FL 34473
Keith Tighe, Director
Joy Nunez, Assistant Director
352-351-7385 Fax: 352-351-7280

Florida Hospital Sleep Disorders Center
601 East Rollins Avenue
Orlando, FL 32803
Morris T. Bird, M.D.
Robert S. Thornton, M.D.
407-897-1558 Fax: 407-897-1775

Orlando Regional Sleep Disorders Center
Orlando Regional Healthcare
  Systems
23 West Copeland Drive
Orlando, FL 32806
Barry Decker, M.D.
Geri Lockhart, BS, RPSGT, RRT
407-649-6869 Fax: 407-872-3876
Email: geril@orhs.org

Health First Sleep Disorders Center
Palm Bay Community Hospital
1425 Malabar Road Northeast,
  Suite 255
Palm Bay, FL 32907
Michael Miller, M.D.
Anna Barker, BA, RPSGT
407-434-8087 Fax: 407-434-8496
Email: abarker@health-first.org

Sleep Disorders Center
West Florida Regional Medical
  Center
8383 North Davis Highway
Pensacola, FL 32514
Jane Wilkinson, Director
David Shaw, M.D., Medical Director
850-494-4850 Fax: 850-494-4809

Sleep Disorders Center
Sarasota Memorial Hospital
1700 South Tamiami Trail
Sarasota, FL 34239
Glenn D. Adams, M.D., Medical
  Director
941-917-2525 Fax: 941-917-6187

St. Petersburg Sleep Disorders Center
2525 Pasadena Avenue South,
  Suite S
St. Petersburg, FL 33707
Neil T. Feldman, M.D.
813-360-0853 and 800-242-3244
  (in Florida)

Tallahassee Sleep Disorders Center
1304 Hodges Drive Suite B
Tallahassee, FL 32308-4613
George F. Slade, M.D.
800-662-4278 x4 or 850-878-7271
Fax: 850-878-1509

Laboratory for Sleep Related Breathing Disorders*
University Community Hospital
3100 East Fletcher Avenue
Tampa, FL 33613
Daniel J. Schwartz, M.D.
Mike Longman, RPSGT, RRT
813-979-7410 Fax: 813-632-7517
Website: www.uch.org

GEORGIA

Sleep Disorders Center
Northside Hospital
5780 Peachtree Dunwoody Road,
  Suite 150
Atlanta, GA 30342
Russell Rosenberg, Ph.D.
John E. Lee, M.D.
David Westerman, M.D.
404-851-8135 Fax: 404-252-9946
Email: nshsleep@mindspring.com

Atlanta Center for Sleep Disorders
303 Parkway Drive, Box 44
Atlanta, GA 30312
Patrick Merrill, RPSGT
Francis Buda, M.D.
Jonne Walter, M.D.
Robert Schnapper, M.D.
404-265-3722 Fax: 404-265-3833

Sleep Disorders Center of Georgia
5505 Peachtree Dunwoody Road,
  Suite 370
Atlanta, GA 30342
D. Alan Lankford, Ph.D.
James J. Wellman, M.D.
404-257-0080 Fax: 404-257-0592

Sleep Disorders Center
Wellstar Cobb Hospital
3950 Austell Road
Austell, GA 30106
Susan T. Keller, Coordinator
Mark Letica, M.D.
Aris Iafridis, M.D., Medical Director
770-732-2250 Fax: 770-732-7217

Central Georgia Sleep Disorders Center
777 Hemlock Street Second Floor
PO Box 1035
Macon, GA 31202
Charles C. Wells, M.D., Medical
  Director
Todd Jones, Technical Director
912-633-7222 Fax: 912-745-5125

Sleep Disorders Center
Wellstar Kennestone Hospital
677 Church Street
Marietta, GA 30060
William Dowdell, M.D.
David Lesch, M.D.
Susan T. Keller, RPSGT
770-793-5353 Fax: 770-793-5357

Savannah Sleep Disorders Center
Saint Joseph's Hospital
6 St. Joseph's Professional Plaza
11706 Mercy Boulevard
Savannah, GA 31419
Anthony M. Costrini, M.D.
912-927-5141 Fax: 912-921-3380
Email: yawn11706@aol.com

Sleep Disorders Center
Memorial Health Systems
4700 Waters Avenue
Savannah, GA 31403
Herbert F. Sanders, M.D.
Stephen L. Morris, M.D.
912-350-8327 Fax: 912-350-7281

Department of Sleep Disorders Medicine
Candler Hospital
5353 Reynolds Street
Savannah, GA 31405
James A. Daly, III, M.D., Medical
  Director
Pamela Rockett, RPSGT, RRT
912-692-6673 Fax: 912-692-6931

HAWAII

Orchid Isle Sleep Disorders Laboratory*
1404 Kilauea Avenue
Hilo, HI 96720
Gilbert J. Ransley, RRT, Technical
  Director
John P. Dawson, M.D., MPH,
  Medical Director
808-935-6105 Fax: 808-935-0016

Pulmonary Sleep Disorders Center*
Kuakini Medical Center
347 North Kuakini Street
Honolulu, HI 96817
Edward J. Morgan, M.D.
Sonia Lee-Gushi, RPSGT, CRTT
808-547-9119 Fax: 808-547-9225
Email: sleephi@juno.com
Website: www.kuakini.org

Sleep Disorders Center of the Pacific
Straub Clinic & Hospital
888 South King Street
Honolulu, HI 96813
James W. Pearce, M.D.
Linda Kapuniai, Dr.P.H.
808-522-4448 Fax: 808-522-3048
Email: sdcop@aloha.net

Queen's Medical Center Sleep Laboratory*
The Queen's Medical Center
1301 Punchbowl Street
Honolulu, HI 96813
Bruce A. G. Soll, M.D.
Jamil Sulieman, M.D.
808-547-4396 Fax: 808-537-7830
Email: bsoll@queens.org

Orchid Isle Sleep Disorders Laboratory*
Waimea Town Plaza
64-1061 Mamalahoa Highway 105
Kamuela, HI 96743
Gilbert J. Ransley, RRT, Technical
  Director
John P. Dawson, M.D., MPH,
  Medical Director
808-885-9681 Fax: 808-885-1705

IDAHO

Idaho Sleep Disorders Center-Boise
St. Luke's Regional Medical Center,
190 East Bannock Street,
Boise, ID 83712
Brett E. Troyer, M.D.
Mary Gable, RPSGT
208-381-2440

Idaho Sleep Disorders Center-Nampa
Mercy Medical Center
1512 12th Avenue Road
Nampa, ID 83686
Brett E. Troyer, M.D.
David K. Merrick
Mary R. Gable
208-463-5820

Idaho Diagnostic Sleep Lab*
526-C Shoup Avenue West
Twin Falls, ID 83301
Ron Fullmer, M.D.
Richard Hammond, M.D.
Brian Fortuin, M.D.
Diana Lincoln-Haye, RRT, RCP
Robin Baggett
208-736-7646 Fax: 208-736-1569

ILLINOIS

Center for Sleep and Ventilatory Disorders
University of Illinois at Chicago
1740 West Taylor Street, M/C 787
Chicago, IL 60612
Robert C. Basner, M.D.
Maureen Smith, RPSGT
312-996-7708 Fax: 312-413-0503
Email: rcbasner@uic.edu

Sleep Disorders Center
The University of Chicago Hospitals
5841 South Maryland, M/C 2091
Chicago, IL 60637
Jean-Paul Spire, M.D.
Wallace B. Mendelson, M.D.,
    Codirector
773-702-1782 Fax: 773-702-7998
Email: eegs@midway.uchicago.edu

Sleep Disorder Service and Research
    Center
Rush-Presbyterian-St. Luke's
    Medical Center
1653 West Congress Parkway
Chicago, IL 60612
Rosalind Cartwright, Ph.D.
312-942-5440 Fax: 312-942-4990
Email: rcartwri@rush.edu

Sleep Disorders Center
Northwestern Memorial Hospital
303 East Superior Passavant 1044
Chicago, IL 60611
Phyllis C. Zee, M.D., Ph.D., Director
James Stockard, M.D., Ph.D.,
    Codirector
312-908-8120 or 312-908-8508
Fax: 312-908-6637
Email: pczee@merle.acns.nw.edu

Sleep Disorders Center
Evanston Hospital
2650 Ridge Avenue
Evanston, IL 60201
Richard S. Rosenberg, Ph.D.
847-570-2567 Fax:847-570-2984
Email: r-rosenberg@nwu.edu

Sleep Disorders Center
Lutheran General Hospital
1775 Dempster Street
Parkside Center, Suite B06
Park Ridge, IL 60068
Barry Weber, M.D.
Wayne Rubinstein, M.D.
Lauren Witcoff, M.D.
847-723-7024 Fax: 847-723-7369

C. Duane Morgan Sleep Disorders Center
Methodist Medical Center of Illinois
221 Northeast Glen Oak Avenue
Peoria, IL 61636
Arthur W. Fox, M.D.
309-672-4966 or 309-671-5136
Fax: 309-672-4117

Sleep Disorders Laboratory*
Rockford Health System
2400 North Rockton Avenue
Rockford, IL 61103
Theodore S. Ingrassia, III, M.D.
815-971-5595 Fax: 815-971-9894

SIU School of Medicine/Memorial Medical Center Sleep Disorders Center
Memorial Medical Center
701 North First Street
Springfield, IL 62781
Joseph Henkle, M.D.
Steven Todd, RRT, RPSGT
217-788-4269 Fax: 217-788-7057

Carle Regional Sleep Disorders Center
Carle Foundation Hospital
611 West Park Street
Urbana, IL 61801-2595
Daniel Picchietti, M.D.
Donald A. Greeley, M.D.
217-383-3364 Fax: 217-383-7117

Sleep Disorders Center
Central Du Page Hospital
25 North Winfield Road
Winfield, IL 60190
Robert Hart, M.D.
Linda Klora, RPSGT
630-682-2975 Fax: 630-682-2745
Email: linda_klora@cdh.org

INDIANA

Sleep Disorders Center
St. Francis Hospital and Health
  Centers
1500 Albany Street, Suite 1110
Beech Grove, IN 46107
Dianna L. Miller, RPSGT
Manfred P. Mueller, M.D., FCCP
317-783-8144 Fax: 317-781-1402

St. Mary's Sleep Disorders Center
St. Mary's Medical Center
3700 Washington Avenue
Evansville, IN 47750
David Cocanower, M.D.
Rebecca N. Dicus
812-485-4960 Fax: 812-485-7953

St. Joseph Sleep Disorders Center
St. Joseph Medical Center
700 Broadway
Fort Wayne, IN 46802
James C. Stevens, M.D.
Thomandram Sekar, M.D.
219-425-3552 Fax: 219-425-3553

Sleep/Wake Disorders Center
Winona Memorial Hospital
3232 North Meridian Street
Indianapolis, IN 46208
Kenneth N. Wiesert, M.D.
317-927-2100 Fax: 317-927-2914

Sleep Disorders Center
St. Vincent Hospital and Health
  Services
8401 Harcourt Road
Indianapolis, IN 46260-0160
Rex McKinney
Thomas Cartwright, M.D.
317-338-2152 Fax: 317-338-4917

Sleep/Wake Disorders Center
Community Hospitals of
  Indianapolis
1500 North Ritter Avenue
Indianapolis, IN 46219
Marvin E. Vollmer, M.D.
317-355-4275 Fax: 317-351-2785

Sleep Alertness Center
Lafayette Home Hospital
2400 South Street
Lafayette, IN 47904
Frederick Robinson, M.D.
765-447-6811 x2840

IOWA

Sleep Disorders Center
  Mary Greeley Medical Center
1111 Duff Avenue
Ames, IA 50010
Selden Spencer, M.D., Director
Mark Hislop, RRT, RPSGT
515-239-2353 Fax: 515-239-6741
Email: sleeplab@mgmc.com

Sleep Disorders Center
The Department of Neurology
The University of Iowa Hospitals
  and Clinics
Iowa City, IA 52242
Mark Eric Dyken, M.D.
319-356-3813 Fax: 319-356-4505
Email: mark-dyken@uiowa.edu

KANSAS

Sleep Disorders Center
Hays Medical Center
201 East 7th Street
Hays, KS 67601
Suzanne Bollig, RRT, RPSGT,
  REEGT
785-623-5373, Fax 785-623-5377
Email: sbollig@haysmed.com

Sleep Disorders Center
St. Francis Hospital and Medical
  Center
1700 Southwest Seventh Street
Topeka, KS 66606-1690
Ted W. Daughety, M.D.
David D. Miller, RPSGT
785-295-7900

Sleep Disorders Center
Wesley Medical Center
550 North Hillside
Wichita, KS 67214-4976
Janice Oeltjenbruns, REEG/EPT,
  RPSGT
Emilio D. Soria, M.D.
316-688-2663 Fax: 316-688-3256

KENTUCKY

Physicians' Center for Sleep Disorders
Graves-Gilbert Clinic
201 Park Street
PO Box 90007
Bowling Green, KY 42102-9007
Michael Zachek, M.D.
Randall Hansbrough, M.D., Ph.D.
Douglas Thomson, M.D., MPH
502-781-5111 Fax: 502-782-4263
Email: zachekm@graves-
  gilbertclinic.com

Sleep Diagnostics Lab*
Greenview Regional Hospital
1801 Ashley Circle
Bowling Green, KY 42101
Steven Zeller, RPSGT
Gul K. Sahetya, M.D.E.
Chandler Deal, M.D.
502-793-2175 Fax: 502-793-2177

Sleep Disorders Center
St. Luke Hospital West
7380 Turfway Road
Florence, KY 41042
Bruce Corser, M.D.
Michael S. Fletcher, RPSGT
606-525-5347 Fax: 606-525-5124
Email: fletcher@healthall.com

The Sleep Disorder Center of St. Luke Hospital
St. Luke Hospital, Inc.
85 North Grand Avenue
Fort Thomas, KY 41075
Bruce Corser, M.D.
Michael S. Fletcher, RPSGT
606-572-3535 Fax: 606-572-3375
Email: fletcher@healthall.com

Sleep Apnea Center*
Jennie Stuart Medical Center
320 West 18th Street
Hopkinsville, KY 42240
Manoj H. Majmudar, M.D.
Mark L. Pierce, RRT, RPSGT
502-887-0410 Fax: 502-887-0412

Sleep Apnea Center*
Samaritan Hospital
310 South Limestone
Lexington, KY 40508
Barbara Phillips, M.D., MSPH, FCCP
Gary King, RRT, Director
606-252-6612 x7331
Fax: 606-252-6612 x7292
Email: bphil95@aol.com

Sleep Disorders Center
St. Joseph's Hospital
One St. Joseph Drive
Lexington, KY 40504
James Thompson, M.D.
Kathryn Hansen, BS
606-278-0444 Fax: 606-260-8021

Sleep Disorders Center
University of Louisville Hospital
530 South Jackson
Street Louisville, KY 40202
Barbara J. Rigdon, RPSGT, R.EEG.T.
Vasudeva G. Iyer, M.D.
Eugene C. Fletcher, M.D.
502-562-3792 Fax: 502-562-4632

Caritas Sleep Apnea Center*
Caritas Medical Center
1850 Bluegrass Avenue
Louisville, KY 40215
Pete Moore, M.D.
William Lacy, M.D.
Richard Baker, M.D.
502-361-6555 Fax: 502-361-6554

Sleep Disorders Center
Audubon Hospital
One Audubon Plaza Drive
Louisville, KY 40217
Pamela McCullough, ARNP
David Winslow, M.D.
502-636-7459 Fax: 502-636-7474

Sleep Medicine Specialists
1169 Eastern Parkway, Suite 3357
Louisville, KY 40217
David H. Winslow, M.D., Director
Darlene R. Herps, RPSGT, Clinical Manager
502-454-0755 Fax: 502-454-3497

Regional Medical Center Lab for Sleep-Related Breathing Disorders*
900 Hospital Drive
Madisonville, KY 42431
Thomas Gallo, M.D.
Frank Taylor, M.D.
502-825-5918 Fax: 502-825-5159

Diller Regional Sleep Disorders Center
Lourdes Hospital
1530 Lone Oak Road
Paducah, KY 42001
James Metcalf, M.D.
Rick Irvan, R.EEG.T., Coordinator
502-444-2660 Fax: 502-444-2661

Breathing Disorders Sleep Lab*
Pikeville Methodist Hospital
911 South Bypass Road
Pikeville, KY 41501
Ramanarao V. Mettu, M.D., FACP, FCCP, Medical Director
Sally Stamper Compton, RRT, Director
Linda Greer, CRTT, Manager
Jerry Miller, Vice-President
606-437-3989 Fax: 606-437-9649

P.A.C. Sleep Disorders Lab*
Pattie A. Clay Hospital
PO Box 1600
801 Eastern Bypass
Richmond, KY 40475
Tom Grant
David Broughton
606-625-3334 Fax: 606-625-3104

The Medical Center Sleep Laboratory*
456 Burnley Road
Scottsville KY 42164
Walter Warren, M.D.
Chris A. Barnett, RRT
502-622-2865 Fax: 502-622-2869

LOUISIANA

Lourdes Sleep Disorders Center
Our Lady of Lourdes Regional
Medical Center
611 St. Landry Street
Lafayette, LA 70506
Robert D Martinez, M.D., Medical
Director
Christine Soileau, RPSGT, R.EEG.T.
318-289-2858 Fax: 318-289-2834

Tulane Sleep Disorders Center
1415 Tulane Avenue
New Orleans, LA 70112
Mark A. McCarthy, M.D.
504-588-5231 Fax: 504-584-1727

Memorial Medical Center Sleep Disorders Center
2700 Napoleon Avenue
New Orleans, LA 70115
Gregory S. Ferriss, M.D.
Li Yu, M.D.
504-896-5652 Fax: 504-896-5772
Email: gferriss@pol.net

LSU Sleep Disorders Center
Louisiana State University Medical
Center
PO Box 33932
Shreveport, LA 71130-3932
Andrew L. Chesson, Jr., M.D.
318-675-5365 Fax: 318-675-4440
Email: achess@lsumc.edu

The Neurology and Sleep Clinic
2205 East 70th Street
Shreveport, LA 71105
Nabil A. Moufarrej, M.D.
Annette Berry, RPFT, RPSGT
318-797-1585 Fax: 318-797-6077
Email: namouf@msn.com

NSRMC Sleep Disorders Center
North Shore Regional Medical
  Center
100 Medical Center Drive
Slidell, LA 70461
Anwant Chawla, M.D.
Mary B. Jones, BS, MT, RPSGT
504-646-5711 Fax: 504-646-5013

MAINE

St. Mary's Sleep Disorders Laboratory*
St. Mary's Regional Medical Center
97 Campus Avenue
Lewiston, ME 04240
Ralph V. Harder, M.D.
Peter J. Leavitt, RRT
207-777-8959

**Maine Institute for Sleep Breathing
  Disorders***
930 Congress Street
Portland, ME 04102
George E. Bokinsky, Jr., M.D.
207-871-4535 Fax: 207-871-6005

MARYLAND

The Johns Hopkins Sleep Disorders Center
Asthma and Allergy Building, Room
  4B50
Johns Hopkins Bayview Medical
  Center
5501 Hopkins Bayview Circle
Baltimore, MD 21224
Philip L. Smith, M.D.,
Alan Schwartz, M.D.
410-550-0571 Fax: 410-550-3374

Maryland Sleep Disorders Center
Greater Baltimore Medical Center
6701 North Charles Street,
  Suite 4140
Baltimore, MD 21204-6808
Thomas E. Hobbins, M.D.
410-494-9773 Fax: 410-823-6635
Email: psrmdteh@igc.apc.org

Frederick Sleep Disorders Center
Frederick Memorial Hospital
400 West Seventh Street
Frederick, MD 21701
Marc Raphaelson, M.D.
Konrad W. Bakker, M.D.
Garland McDonald
301-698-3802

**The Sleep-Breathing Disorders Center of
  Hagerstown***
12821 Oak Hill Avenue
Hagerstown, MD 21742
Abdul Waheed, M.D.
Shaheen Iqbal, M.D.
Johny Alencherry, M.D.
301-733-5971 Fax: 301-733-5773

Shady Grove Sleep Disorders Center
14915 Broschart Road, Suite 102
Rockville, MD 20850
Jean Neuenkirch, RPSGT
301-251-5905 Fax: 301-251-6189

**Washington Adventist Sleep Disorders
  Center**
7525 Carroll Avenue
Takoma Park, MD 20912
Marc Raphaelson, M.D.
Konrad W. Bakker, M.D.
301-891-2594

MASSACHUSETTS

Sleep Disorders Center
Beth Israel Deaconess Medical
  Center
330 Brookline Avenue, KS430
Boston, MA 02215
Jean K. Matheson, M.D.
Janet Mullington, Ph.D.
617-667-3237 Fax: 617-975-5506
Email: jmatheso@
  bidmc.harvard.edu

Sleep Disorders Center
Lahey Clinic
41 Mall Road
Burlington, MA 01805
Paul T. Gross, M.D.
Susan M. Dignan, RPSGT
781-744-8251 Fax: 781-744-5243
Email: susan.m.dignan@lahey.org

**Sleep Disorders Institute of Central New
  England**
St. Vincent Hospital
25 Winthrop Street
Worcester, MA 01604
Jayant G. Phadke, M.D.
508-798-6212 Fax: 508-798-6373

MICHIGAN

Sleep Disorders Center
University of Michigan Hospitals
1500 East Medical Center Drive,
  UH8D 8702
Box 0117
Ann Arbor, MI 48109-0115
Brenda Livingston, Coordinator
Michael S. Aldrich, M.D.
Ronald Chervin, M.D.
Beth Malow, M.D.
734-936-9068 Fax: 734-936-5377

Sleep Disorders Center
St. Joseph Mercy Hospital
PO Box 995
Ann Arbor, MI 48106
Thomas R. Gravelyn, M.D.
Sharon S. Potoczak, RPSGT
734-712-4651

Sleep Disorders Clinic
Bay Medical Center
1900 Columbus Avenue
Bay City, MI 48708
John M. Buday, M.D.
Mary K. Taylor, RPSGT, RRT
517-894-3332 Fax: 517-894-6114

Harper Hospital Sleep Disorders Center
4160 John R, Suite 400
Detroit, MI 48201
James A. Rowley, M.D.
313-745-9009 Fax:313-745-8725

Sleep/Wake Disorders Laboratory (127B)
VA Medical Center
4646 John R. Street
Detroit, MI 48201-1916
Sheldon Kapen, M.D.
M. Safwan Badr, M.D.
Greg Koshorek
313-576-3663 Fax: 313-576-1122
Email: kapen.sheldon@
allenpark.va.gov

Sinai Sleep Center DMC
Sinai Hospital
6767 West Outer Drive
Detroit, MI 48235-2899
Bradley Rowens, M.D.
Keith Williams
313-493-5148 Fax: 313-493-5036

West Michigan Sleep Disorders Center
Butterworth Hospital
100 Michigan Street Northeast
Grand Rapids, MI 49503
Lee Marmion, M.D.
Ronald Van Drunen, RPSGT
616-391-3759 Fax: 616-391-3052

Sleep Disorders Center
Borgess Medical Center
1521 Gull Road
Kalamazoo, MI 49001
Sue Cammarata, M.D.
Thomas Wittenberg, RRT
Sheri Dillon, RRT
616-226-7081 Fax: 616-226-6909

Ingham Regional Medical Center
Sleep/Wake Center
2025 South Washington Avenue,
Suite 300
Lansing, MI 48910-0817
Pamela Minkley, RRT, RPSGT
Gauresh Kashyap, M.D., FACP,
FCCP
517-372-6444 Fax: 517-372-6440
Email: pminkley@juno.com

Sparrow Sleep Center
Sparrow Hospital
1215 East Michigan Avenue
PO Box 30480
Lansing, MI 48909-7980
Alan M. Atkinson, D.O.
David K. Young, D.O.
517-364-5370 Fax: 517-364-5373

Sleep & Respiratory Associates of Michigan
28200 Franklin Road
Southfield, MI 48034
Harvey W. Organek, M.D.
248-350-2722 Fax: 248-350-0154

Munson Sleep Disorders Center
Munson Medical Center
1105 Sixth Street MPB, Suite 307
Traverse City, MI 49684-2386
David A. Walker, D.O., FCCP,
Medical Director
Leon R. Olewinski, RRT, Director
Marcia Rinal, CRTT, RPSGT,
Manager
800-358-9641 and 616-935-6600
Fax: 616-935-6610

Sleep Disorders Institute
44199 Dequindre, Suite 311
Troy, MI 48098
R. Bart Sangal, M.D.
248-879-0707 Fax: 248-879-2704
Email: sangalrb@mindspring.com

MINNESOTA

Duluth Regional Sleep Disorders Center
St. Mary's Duluth Clinic Health
  System
407 East Third Street
Duluth, MN 55805
Peter K. Franklin, M.D.
Paul J. Windberg, M.D.
Mary Carlson, RPSGT
218-726-4692 Fax: 218-726-4083

Fairview Sleep Center
Fairview Southdale Hospital
6401 France Avenue
South Edina, MN 55435
John E. Trusheim, M.D.
Eileen M. Peterson
612-924-5053 Fax: 612-924-5994
Email: epeters1@fairview.org

Minnesota Regional Sleep Disorders
  Center #867B
Hennepin County Medical Center
701 Park Avenue South
Minneapolis, MN 55415
Mark Mahowald, M.D.
612-347-6288 Fax: 612-904-4207
Email: mahow002@
  maroon.tc.umn.edu

Sleep Disorders Center
Abbott Northwest Hospital
800 East 28th Street
Minneapolis, MN 55407
Wilfred A. Corson, M.D.
612-863-4516 Fax: 612-863-2837

Mayo Sleep Disorders Center
Mayo Clinic
200 First Street Southwest
Rochester, MN 55905
Peter Hauri, Ph.D.
John W. Shepard, Jr., M.D.
507-266-8900 Fax: 507-266-7772
Email: hauri.peter@mayo.edu

Sleep Disorders Center
Methodist Hospital
6500 Excelsior Boulevard
St. Louis Park, MN 55426
Barb Feider, RPSGT
Salim Kathawalla, M.D.
612-993-6083 Fax: 612-993-7026

St. Joseph's Sleep Diagnostic Center
St. Joseph's Hospital
69 West Exchange Street
St. Paul, MN 55102
Thomas Mulrooney, M.D.
612-232-3682 Fax: 612-232-4111

MISSISSIPPI

Sleep Disorders Center
Memorial Hospital at Gulfport
PO Box 1810
Gulfport, MS 39501
Sydney Smith, M.D.
601-865-3152 Fax: 601-865-3259

Sleep Disorders Center
Forrest General Hospital
6051 Highway 49
PO Box 16389
Hattiesburg, MS 39404-6389
Geoffrey B. Hartwig, M.D.
John R. Harsh, Ph.D.
Dennis Kramer
601-288-4790 and 800-280-8520
Fax: 601-288-4791
Email: fghsdc@netdoor.com
Website: www.forrestgeneral.com/
   sleepdisorders.htm

Sleep Disorders Center
University of Mississippi Medical
   Center
2500 North State Street
Jackson, MS 39216-4505
Howard Roffwarg, M.D., Director
Alp Sinan Baran, M.D., Medical
   Director
Allen Richert, M.D., Staff Specialist
601-984-4820 Fax: 601-984-5885
Email: asbaran@pol.net

MISSOURI

Unity Sleep Medicine and Research Center
St. Luke's Hospital
232 South Woods Mill Road
Chesterfield, MO 63017
James K. Walsh, Ph.D.
Gihan Kader, M.D.
314-205-6030 Fax: 314-205-6025
Email: jkw@stlo.smhs.com and
   gak@stlo.smhs.com

University of Missouri Sleep Disorders
   Center University Hospital and Clinics
M-741 Neurology
One Hospital Drive
Columbia, MO 65212
Pradeep Sahota, M.D.
573-884-SLEEP and 800-ADD-
   SLEEP
Fax: 573-884-4785
Email: sahotp@brain.missouri.edu

Sleep Disorders Center
Research Medical Center
2316 East Meyer Boulevard
Kansas City, MO 64132-1199
Jon D. Magee, Ph.D.
816-276-4334 Fax: 816-276-3488

Sleep Disorders Center
St. Luke's Hospital
4400 Wornall Road
Kansas City, MO 64111
Ann Romaker, M.D.
Wendy L. Fluegel, M.D.
816-932-3207 Fax: 816-932-3383

Cox Regional Sleep Disorders Center
3800 South National Avenue,
   Suite LL
150 Springfield, MO 65807
Edward Gwin, M.D.
417-269-5575 Fax: 417-269-5578

St. John's Sleep Disorders Center
St. John's Regional Health Center
1235 East Cherokee
Springfield, MO 65804
John Brabson, M.D., Medical
   Director
Terry M. Yarnell, REEGT, RCPT,
   Administrative Director
417-885-5464 Fax: 417-885-5465

Sleep Disorders & Research Center
Deaconess Medical Center
6150 Oakland Avenue
St. Louis, MO 63139
Sidney D. Nau, Ph.D.
Korgi V. Hegde, M.D.
314-768-3100 Fax: 314-768-3594

Sleep/Wake Disorders Center SLU Care
The Health Services Division of
    Saint Louis University
1221 South Grand Boulevard
St. Louis, MO 63104
Shashidhar M. Shettar, M.D.
314-577-8705 Fax: 314-664-7248
Email: shettars@wpogate.slu.edu

MONTANA

The Sleep Center at St. Vincent Hospital
St. Vincent Hospital and Health
    Center
1233 North 30th Street
Billings, MT 59101
William C. Kohler, M.D.
Karen Y. Allen, CRTT, RPSGT
406-238-6815 Fax: 406-238-6262
Email: wkohler@svhhc.org or
    kallen@svhhc.org

Sleep Disorders Center
Deaconess Billings Clinic
2800 Tenth Avenue North
PO Box 37000
Billings, MT 59107
Rich Lundy, BS, RRT
Robert K. Merchant, M.D.
406-657-4075 Fax: 406-657-4717
Email: rlundy@billingsclinic.org
Website: www.billingsclinic.org

NEBRASKA

Adult and Pediatric Sleep Related
    Breathing Disorders Laboratory*
Bryan LGH Memorial Center East
1600 South 48th Street
Lincoln, NE 68506
Debra Bailey, RN, Clinical Manager
Jack Mathews, M.D., Medical
    Director
402-483-3950 Fax: 402-483-8374
Email: dbailey@bryanlgh.org

Great Plains Regional Sleep Physiology
    Center
Bryan LGH Medical Center West
2300 South 16th Street
Lincoln, NE 68502
Timothy R. Lieske, M.D.
Leigh Heithoff, RPSGT, R.EEG.T.
402-473-5338 Fax:402-473-5380
Email: laheitho@lgh.org

Sleep Disorders Center
Nebraska Health System 987546
Nebraska Medical Center
Omaha, NE 68198-7546
Carie L. Smith, RRT, RPSGT
Stephen B. Smith, M.D.
402-552-2286 Fax: 402-552-2057

Sleep Disorders Center
Methodist/Richard Young Hospital
2566 St. Mary's Avenue
Omaha, NE 68105
Robert J. Ellingson, Ph.D., M.D.
John D. Roehrs, M.D.
402-354-6305 and 402-354-6309
Fax: 402-354-6334

NEVADA

Mountain Medical Sleep Disorders Center
Mountain Medical Associates,Inc.
710 West Washington Street
Carson City, NV 89703-3826
Robert L. McDonald, M.D.
John Zimmerman, Ph.D.
702-882-2106 Fax 702-882-0838
email: dzimmer889@aol.com

The Sleep Clinic of Nevada
1012 East Sahara Avenue
Las Vegas, NV 89104
Darlene Steljes, CEO
702-893-0020 Fax: 702-893-0025
Email: yingyang@ix.netcom.com

Washoe Sleep Disorders Center and Sleep Laboratory
Washoe Professional Building and Washoe Medical Center
Sleep Management, Inc.
75 Pringle Way, Suite 701
Reno, NV 89502
William C. Torch, M.D., MS
John T. Zimmerman, Ph.D.
Kathleen Auld, D.O.
702-328-4700 and 800-JETLAGG
Fax: 702-329-2715

NEW HAMPSHIRE

Sleep Disorders Center
Dartmouth-Hitchcock Medical Center
One Medical Center Drive
Lebanon, NH 03756
Michael Sateia, M.D.
603-650-7534 Fax: 603-650-7820
Email: sleep.disorders.center@ dartmouth.edu

Center for Sleep Evaluation
Catholic Medical Center
100 McGregor Street
Manchester, NH 03102
Peter E. Corrigan, M.D., Medical Director
Jeanetta C. Rains, Ph.D., Clinical Director
603-663-6680 Fax: 603-663-6699

NEW JERSEY

SleepCare Center of Cherry Hill
457 Haddonfield Road, Suite 520
Cherry Hill, NJ 08002
James La Russo, Chief Executive Officer
John D.Miladin, President/Chief Operating Officer
Kathleen L. Ryan, M.D., FCCP, FACP
800-753-3779 Fax: 609-662-5187
Email: flajack@aol.com
Website: www.sleepcarecenter.com

Sleep Lab Institute for Sleep/Wake Disorders
Hackensack University Medical Center
30 Prospect Avenue
Hackensack, NJ 07601
Hormoz Ashtyani, M.D.
Sue Zafarlotfi, Ph.D.
201-996-2992

Morristown Sleep Disorder Center
Morristown Memorial Hospital
95 Mount Kemble Avenue
Morristown, NJ 07962
Robert A. Capone, M.D., FCCP
Pamela Wolfsie, RPSGT
973-971-4567 Fax: 973-290-7620
Website: www.atlantichealth.org

SleepCare Memorial Hospital of Burlington County
175 Madison Avenue
Mount Holly, NJ 08060
Kathleen L. Ryan, M.D., FCCP, FACP
Jack Miladin
800-753-3779 Fax: 609-662-5187
Email: flajack@aol.com
Website: www.sleepcarecenter.com

Comprehensive Sleep Disorders Center
Robert Wood Johnson University
   Hospital/ UMDNJ
Robert Wood Johnson Medical
   School
One Robert Wood Johnson Place
PO Box 2601
New Brunswick, NJ 08903-2601
Richard A. Parisi, M.D.
Raymond Rosen, Ph.D.
732-937-8683 Fax: 732-418-8448

Sleep Disorders Center
Newark Beth Israel Medical Center
201 Lyons Avenue
Newark, NJ 07112
Monroe Karetzky, M.D.
973-926-6668 Fax: 973-923-6672

Sleep Disorders Center
Capital Health System at Mercer
446 Bellevue Avenue
PO Box 1658
Trenton, NJ 08607
Debra DeLuca, M.D.
Rita Brooks, R.EEG/EP.T., RPSGT,
   CNIM
Paula Page
609-394-4167 Fax: 609-394-4352

Snoring and Sleep Apnea Center*
Helene Fuld Medical Center
750 Brunswick Avenue
Trenton, NJ 08638
Marcella Frank, D.O.
Rita Brooks, R.EEG/EP.T., RPSGT,
   CNIM
609-278-6990 Fax: 609-278-6982

Sleep Disorders Center of New Jersey
2253 South Avenue, Suite 7
Westfield, NJ 07090
David S. Goldstein, M.D.
Michael Lahey, RPSGT
908-789-4244 Fax: 908-789-2716

NEW MEXICO

University Hospital Sleep Disorders Center
4775 Indian School Road Northeast,
   Suite 307
Albuquerque, NM 87110
Rose Mills Barry Krakow, M.D.
505-272-6101 Fax: 505-272-6112

Lovelace Sleep Disorders Center
Lovelace Health Systems
2929 Coors Boulevard NW,
  Suite 106
Albuquerque, NM 87120
Lee K. Brown, M.D., Medical
  Director
Nancy L. Polnaszek, Director of
  Medical Specialties
505-839-2369 Fax: 505-839-2378
Email: nlpoln@lovelace.com
Website: www.lovelace.com

NEW YORK

**Capital Region Sleep/Wake Disorders
  Center**
St. Peter's Hospital
Pine West Plaza #1
Washington Avenue Extension
Albany, NY 12205
Aaron E. Sher, M.D.
Paul B. Glovinsky, Ph.D.
518-436-9253

Sleep/Wake Disorders Center
Montefiore Medical Center
111 East 210th Street
Bronx, NY 10467
Michael J. Thorpy, M.D.
718-920-4841 Fax: 718-798-4352
Email: thorpy@aecom.yu.edu

Bassett Healthcare Sleep Disorders Center
Bassett Healthcare
One Atwell Road
Cooperstown, NY 13326
Lee C. Edmonds, M.D.
Robert C. Reese, RRT, RPSGT
607-547-6979 Fax: 607-547-6995
Email: reesere@hotmail.com
Website: www.bassetthealthcare.org

**St. Joseph's Hospital Sleep Disorders
  Center**
St. Joseph's Hospital
555 East Market Street
Elmira, NY 14902
Kathleen R. Reilly, BS, RRT
Paula Cook, RPSGT, RRT
607-737-7008 Fax: 607-737-1522

Sleep Disorders Center
Winthrop-University Hospital
222 Station Plaza North
Mineola, NY 11501
Michael Weinstein, M.D., FCCP
Maritza Groth, M.D., FCCP
Claude Albertario, RPSGT
516-663-3907 Fax:516-663-4788
Email: mweinstein@winthrop.org

Sleep-Wake Disorders Center
Long Island Jewish Medical Center
270-05 76th Avenue
New Hyde Park, NY 11042
Harly Greenberg, M.D.
Jane Luchsinger, MS
718-470-7058 Fax: 718-470-7058
Email: greenber@lij.edu

The Sleep Disorders Center
Columbia-Presbyterian Medical
   Center
161 Fort Washington Avenue
New York, NY 10032
Neil B. Kavey, M.D.
212-305-1860 and 914-948-0400
Fax: 212-305-5496
Email: nbk1@columbia.edu

Sleep Disorders Institute
1090 Amsterdam Avenue
New York, NY 10025
Gary K. Zammit, Ph.D.
212-523-1700 Fax: 212-523-1704
Email: gzammit@slrhc.org

Sleep-Wake Disorders Center
New York Hospital-Cornell
   Manhattan Campus
520 East 70th Street
New York, NY 10021
Daniel Wagner, M.D.
Margaret Moline, Ph.D.
914-997-5751

Sleep Disorders Center of Rochester
2110 Clinton Avenue South
Rochester, NY 14618
Donald W. Greenblatt, M.D.
716-442-4141 Fax:716-442-6259

Sleep Disorders Center
State University of New York at
   Stony Brook
University Hospital, MR 120 A
Stony Brook, NY 11794-7139
Marta Maczaj, M.D.
516-444-2916 Fax: 516-444-7851

The Sleep Center
Community General Hospital
Broad Road
Syracuse, NY 13215
Robert E. Westlake, M.D.
Bruce D. Hall, RPSGT, RRT
315-492-5877 Fax: 315-492-5521
Website: www.cgh.org

The Sleep Laboratory*
St. Joseph's Hospital Health Center
945 East Genesee Street Suite 300
Syracuse, NY 13210
Edward T. Downing, M.D.
Stephen F. Swierczek, RPSGT
315-475-3379 Fax: 315-475-5077
Website: www.sjhsyr.org

The Mohawk Valley Sleep Disorders Center
St. Elizabeth Medical Center
2209 Genesee Street
Utica, NY 13501
Steven A. Levine, D.O., FCCP
Mark Cassidy, RPSGT
315-734-3484 Fax: 315-734-3494
Email: mvsdc@stemc.org

Sleep-Wake Disorders Center
The New York Hospital-Cornell
   Medical Center
21 Bloomingdale Road
White Plains, NY 10605
Daniel R. Wagner, M.D. Medical
   Director
Margaret L. Moline, Ph.D., Director
914-997-5751 Fax: 914-682-6911
Email:
   dwagner%westnyh@nyh.med.
   cornell.edu
or mmoline%westnyh@nyh.med.
   cornell.edu

The Sleep Disorders Center
White Plains Columbia-Presbyterian
  Medical Center
185 Maple Avenue
White Plains, NY 10601
Neil B. Kavey, M.D.
914-948-0400 Fax: 212-305-5496
Email: nbk1@columbia.edu

NORTH CAROLINA

Western Carolina Sleep Center
Mission/St. Joseph's Health System
445 Biltmore Avenue, Suite 404
Asheville, NC 28801
Charles O'Cain, M.D.
James McCarrick, M.D.
Jean C. Hardy, RPSGT
828-258-6701 Fax: 828-258-6702

Sleep Medicine Center of WNC
1091 Hendersonville Road
Asheville, NC 28803
John S. Morris, M.D.
Harriet Pruitt, RPSGT
704-277-7533 Fax: 704-277-7493

Carolinas Sleep Services
Mercy Hospital South
16028 Park Road
Charlotte, NC 28210
Mindy Beth Cetel, M.D.
Michael Stolzenbach
704-543-2213

Carolinas Sleep Services
University Hospital
8800 North Tyron Street
PO Box 560727
Charlotte, NC 28256
Mindy B. Cetel, M.D.
Paul D. Knowles, M.D.
Mary Susan Esther, M.D.
Michael Stolzenbach, RPSGT
704-548-5855 Fax: 704-548-6848

Sleep Disorders Center
Moses Cone Health System
1200 North Elm Street
Greensboro, NC 27401-1020
Clinton D. Young, M.D.
Reggie Whitsett, RPSGT
336-832-7406 Fax: 336-832-8649

Sleep Medicine Center of Salisbury
911 West Henderson Street,
  Suite L30
Salisbury, NC 28144
Dennis L. Hill, M.D.
Deborah Kooy, RPSGT
704-637-1533 Fax: 704-637-0470
Email: drhill@salisbury.net

Sleep Disorders Center
North Carolina Baptist Hospital
Wake Forest University School of
  Medicine
Medical Center Boulevard
Winston-Salem, NC 27157
W. Vaughn McCall, M.D.
Linda Quinlivan
336-716-5288
Email: vmcall@wfubmc.edu

Summit Sleep Disorders Center
160 Charlois Boulevard
Winston-Salem, NC 27103
J. Baldwin Smith, III, M.D.
Richard Doud Bey, M.D.
336-765-9431 Fax: 336-765-4889
Email: minnie@netunlimited.net

NORTH DAKOTA

No accredited center members.

OHIO

The Tri-State Sleep Disorders Center
1275 East Kemper Road
Cincinnati, OH 45246
Martin B. Scharf, Ph.D.
513-671-3101 Fax:513-671-4159
Email: sleepsatl@aol.com
Website: www.trihealth.org

Cincinnati Regional Sleep Centers
2123 Auburn Avenue, Suite 322
Cincinnati, OH 45219
Jospeh W. Zompero, RPSGT
Bruce C. Corser, M.D.
513-721-4680 Fax: 513-721-1036

Sleep Disorders Center of Greater Cincinnati
TriHealth Hospitals
619 Oak Street
Cincinnati, OH 45206
Virgil Wooten, M.D.
513-569-6320 Fax: 513-569-5495
Email: virgil_wooten@trihealth.com

Sleep Disorders Center
The Cleveland Clinic Foundation
9500 Euclid Avenue, Desk S-51
Cleveland, OH 44195
Dudley S. Dinner, M.D.
216-444-2165 Fax: 216-445-4378

PMA Cardiopulmonary Sleep Laboratory*
Pulmonary Medicine Associates, Inc.
15805 Puritas Avenue
Cleveland, OH 44135
Paul C. Venizelos, M.D., FCCP
Babu M. Eapen, M.D., FCCP
Belinda Gray, RPSGT
216-267-5933 Fax: 216-267-5133

University Hospitals Sleep Center
University Hospitals of Cleveland
Department of Neurology
11100 Euclid Avenue
Cleveland, OH 44106
Carl Rosenberg, M.D., MBA
Lucica Buzoianu
216-844-1301 Fax:216-844-8753

Sleep Disorders Center
The Ohio State University Medical Center
Rhodes Hall, S1039
410 West 10th Avenue
Columbus, OH 43210-1228
Greg Landholt, RPSGT
Charles P. Pollak, M.D.
614-293-8296 Fax:614-293-4506

Good Samaritan Sleep Lab*
Good Samaritan Hospital
2222 Philadelphia Drive
Dayton, OH 45406
Joyce E. Gray, Manager
937-276-8307 Fax: 937-276-8228

The Center for Sleep & Wake Disorders
Miami Valley Hospital
One Wyoming Street, Suite G-200
Dayton, OH 45409
James Graham, M.D.
Kevin Huban, Psy.D.
937-208-2515

Ohio Sleep Medicine and Neuroscience Institute
4975 Bradenton Avenue
Dublin, OH 43017
Helmut S. Schmidt, M.D.
Betty Hammonds
614-766-0773 Fax: 614-766-2599
Email: sleepohio@aol.com

Sleep Disorders Center
Kettering Medical Center
3535 Southern Boulevard
Kettering, OH 45429-1295
Donna Arand, Ph.D.
937-296-7805 Fax: 937-296-7821
Email: donna_arand@
   ketthealth.com

Ohio Sleep Disorders Center
150 Springside Drive
Montrose, OH 44333
Jose Rafecas, M.D.
Frankie Roman, M.D.
330-670-1290 Fax: 330-670-1292

Northwest Ohio Sleep Disorders Center
The Toledo Hospital
Harris-McIntosh Tower, Second
   Floor
2142 North Cove Boulevard
Toledo, OH 43606
Pam Lang, RPSGT
Frank O. Horton, III, M.D.
419-471-5629 Fax: 419-479-6954

Sleep Disorders Center
Genesis Health Care System
Good Samaritan Medical Center
800 Forest Avenue
Zanesville, OH 43701
Roger J. Balogh, M.D.
Thomas E. Rojewski, M.D.
Robert J. Thompson, M.D.
740-454-5855 Fax: 740-455-7646

OKLAHOMA

Sleep Disorders Center of Oklahoma
Integris Health
4401 South Western Avenue
Oklahoma City, OK 73109
Jonathan R.L. Schwartz, M.D.
Elliott R. Schwartz, D.O.
Chris A. Veit, M.S.W., RPSGT
405-636-7700 Fax: 405-636-7531
Email: veitca@integris-health.com
Website: www.integris-health.com

OREGON

Sleep Disorders Center
Sacred Heart Medical Center
1255 Hilyard Street
PO Box 10905
Eugene, OR 97440
Rodney Roth, RRT, RCP
Robert Tearse, M.D.
503-686-7224

Sleep Disorders Center
Rogue Valley Medical Center
2825 East Barnett Road
Medford, OR 97504
Eric Overland, M.D.
Michael Schwartz, RPSGT
Nic Butkov, RPSGT
541-608-4320 Fax: 541-608-5890

Pacific NW Sleep/Wake Disorders Program
Suite 202 1849 NW Kearney
Portland, OR 97209
Gerald B. Rich, M.D.
Ranae Beck
503-228-4414 Fax: 503-228-7293

Sleep Disorders Laboratory*
Providence Portland Medical Center
4805 Northeast Glisan Street
Portland, OR 97213
Louis Libby, M.D.
Keith D. Hyde, MBA, RRT
Dianne Hurst, CRET
503-215-6552 Fax: 503-215-6031

Legacy Good Samaritan Sleep Disorders Center
Neurology, T-302
1015 Northwest 22nd Avenue
Portland, OR 97210
John J. Greve, M.D., Medical Director,
Jan White, Manager
503-413-7540 Fax: 503-413-6919

Salem Hospital Sleep Disorders Center
Salem Hospital
665 Winter Street
Southeast Salem, OR 97309-5014
Mark T. Gabr, M.D.
Stephen J. Baughman, RRT, RPSGT
503-370-5170 Fax: 503-375-4722
Email: sjbaug@salemhospital.org

PENNSYLVANIA

Sleep Disorders Center
Abington Memorial Hospital
1200 Old York Road
Rorer Building, 2nd Floor
Abington, PA 19001
B. Franklin Diamond, M.D.
Albert D. Wagman, M.D.
Kevin R. Booth, M.D.
215-576-2226 Fax: 215-576-2730
Email: bfd13042@home.com
Website: www.amh.org

Sacred Heart Sleep Disorders Center
Sacred Heart Hospital
421 Chew Street
Allentown, PA 18102-3490
William R. Pistone, D.O.
Ross Futerfas, M.D.
K. Alexander Haraldsted, M.D.
David J. Brooks, RRT, RPSGT
610-776-5333 Fax: 610-776-5110
Email: shh_sleep@juno.com

Sleep Disorders Center
Lower Bucks Hospital
501 Bath Road
Bristol, PA 19007
Howard J. Lee, M.D.
215-785-9752 Fax: 215-785-9068

Penn Center for Sleep Disorders
800 West State Street
Doylestown, PA 18901
Richard J. Schwab, M.D.
Allan I. Pack, M.D., Ph.D.
Louis Metzger
215-345-5003 Fax: 215-345-5047

Sleep Disorders Center of Lancaster
Lancaster General Hospital
555 North Duke Street
Lancaster, PA 17604-3555
Harshadkumar B. Patel, M.D.
James M. O'Connor, RPSGT
717-290-5910 Fax: 717-290-4964

Saint Mary Sleep/Wake Disorder Center
Langhorne-Newtown Road
Langhorne, PA 19047
Howard J. Lee, M.D., Medical
   Director
James J. Burke, Administrative
   Director
215-741-6744 Fax: 215-741-6695

Sleep Medicine Services
Paoli Memorial Hospital
255 West Lancaster Avenue
Paoli, PA 19301
Donald D. Peterson, M.D.
Mark R. Pressman, Ph.D.
610-645-3400 Fax: 610-645-2291
Email: pressman@mlhs.org

**Pennsylvania Hospital Sleep Disorders
   Center**
Pennsylvania Hospital
Eighth and Spruce Streets
Philadelphia, PA 19107
Charles R. Cantor, M.D.
Ronald L. Kotler, M.D.
215-829-7079 Fax: 215-829-5630

Penn Center for Sleep Disorders
University of Pennsylvania Medical
   Center
3400 Spruce Street, 11 Gates West
Philadelphia, PA 19104
Allan I. Pack, M.D., Ph.D.
Richard J. Schwab, M.D.
Louis F. Metzger
215-662-7772 Fax: 215-349-8038
Email:
   loumetzg@mail.med.upenn.edu

Temple Sleep Disorders Center
Temple University Hospital
3401 North Broad Street
Rock Pavilion, 4th Floor
Philadelphia, PA 19140
Samuel Krachman, D.O.
Grace R. Denault, BA, RPSGT
215-707-8163 Fax: 215-707-3876

**Sleep Disorders Center, Department of
   Neurology MCP Hahnemann School of
   Medicine**
Allegheny University of the Health
   Sciences
3200 Henry Avenue
Philadelphia, PA 19129
June M. Fry, M.D., Ph.D., Chief,
   Division of Somnology
215-842-4250 Fax: 215-848-3850

Sleep Disorders Center
Thomas Jefferson University
1015 Walnut Street, Suite 319
Philadelphia, PA 19107
Karl Doghramji, M.D.
215-955-6175 Fax: 215-959-9783
Email: karl.doghramji@mail.tju.edu

Sleep and Chronobiology Center
Western Psychiatric Institute and
  Clinic
3811 O'Hara Street
Pittsburgh, PA 15213-2593
Charles F. Reynolds III, M.D.
412-624-2246 Fax: 412-624-2841

Pulmonary Sleep Evaluation Laboratory*
University of Pittsburgh Medical
  Center
Montefiore University Hospital
3459 Fifth Avenue, S639
Pittsburgh, PA 15213
Nancy Kern, CRTT, RPSGT
Mark H. Sanders, M.D.
Patrick J. Strollo, M.D.
412-692-2880 Fax: 412-692-2888

Crozer Sleep Disorders Center
Taylor Hospital
175 East Chester Pike
Ridley Park, PA 19078
Calvin Stafford, M.D., Clinical
  Director
610-595-6272 Fax: 610-595-6273

Sleep Disorders Center
Community Medical Center
1822 Mulberry Street
Scranton, PA 18510
S. Ramakrishna, M.D., FCCP
717-969-8931

Sleep Disorders Center
Mercy Hospital
25 Church Street
Wilkes-Barre PA 18765
John Della Rosa, M.D.
717-826-3410 Fax: 717-820-6658

Sleep Disorders Center
The Lankenau Hospital
100 Lancaster Avenue
Wynnewood, PA 19096
Mark R. Pressman, Ph.D.
Donald D. Peterson, M.D.
610-645-3400

RHODE ISLAND

No accredited center members.

SOUTH CAROLINA

Roper Sleep/Wake Disorders Center
Roper Hospital
316 Calhoun Street
Charleston, SC 29401-1125
William T. Dawson, Jr., M.D.
Wayne C. Vial, M.D.
Graham C. Scott, M.D.
John A. Mitchell, M.D.
Tim Fultz, MS, RRT, RPSGT
843-724-2246 Fax: 843-724-2765
Email: tim.fultz@carealliance.com
Website: www.carealliance.com

Sleep Disorders Center of
South Carolina Baptist Medical
  Center
Taylor at Marion Streets
Columbia, SC 29220
Richard Bogan, M.D., FCCP
Sharon S. Ellis, M.D., Neonatologist
803-771-5847 or 800-368-1971
Fax: 803-401-3080

Southeast Regional Sleep Disorders Center-Easley
200 Fleetwood Drive
PO Box 2129
Easley, SC 29640
Freddie E. Wilson, M.D., Medical
   Director
Katrinka Scalise
864-855-7200 Fax: 864-627-9301

Southeast Regional Sleep Disorders Center
3900 Pelham Road
Greenville, SC 29615
Freddie E. Wilson, M.D., Medical
   Director
Cathy DeJong, RRT, RPSGT, RCP,
   Clinical Manager
Katrinka Scalise, Facility Manager
864-627-5337 Fax: 864-627-9301

Sleep Disorders Center
Greenville Memorial Hospital
701 Grove Road
Greenville, SC 29605
Don McMahan
864-455-8916 Fax: 864-455-4670

Carolinas Sleep Services
1665 Herlong Court, Suite B
Rock Hill, SC 29732
Michael A. Stolzenbach, RPSGT,
   Manager
William C. Sherrill, M.D., Medical
   Director
803-817-1915

Sleep Disorders Center
Spartanburg Regional Medical
   Center
101 East Wood Street
Spartanburg, SC 29303
Shari Angel Newman, RPSGT
864-560-6904 Fax: 864-560-7083

SOUTH DAKOTA

The Sleep Center
Rapid City Regional Hospital
353 Fairmont Boulevard
PO Box 6000
Rapid City, SD 57709
K. Alan Kelts, M.D., Ph.D.
Terry Anderson, BS, RRCP
605-341-8037

Sleep Disorders Center
Sioux Valley Hospital
1100 South Euclid
Sioux Falls, SD 57117-5039
Liz Grav
605-333-6302 Fax:605-333-4402
Email: gravl@siouxvalley.org

TENNESSEE

Summit Center for Sleep Related Breathing Disorders*
Columbia-Summit Medical Center
5655 Frist Boulevard MOB,
   Suite 401
Hermitage, TN 37076
Timothy L. Morgenthaler, M.D.
Lee Ann Covington, RRT, RPSGT
615-316-3495 Fax: 615-316-3493

Sleep Disorders Laboratory*
Regional Hospital of Jackson
367 Hospital Boulevard
Jackson, TN 38303
Thomas W. Ellis, M.D.
David M. Larsen, M.D.
Charlie Carroll, RPSGT
901-661-2148 Fax: 901-661-2441

Sleep Disorders Center
Ft. Sanders Regional Medical Center
1901 West Clinch Avenue
Knoxville, TN 37916
Thomas G. Higgins, M.D.
Bert A. Hampton, M.D.
C. Keith Hulse, Ph.D.
423-541-1375 Fax: 423-541-1714

Sleep Disorders Center
St. Mary's Medical Center
900 East Oak Hill Avenue
Knoxville, TN 37917-4556
William Finley, Ph.D., Director
423-545-6746 Fax: 423-545-3115
Email: bfinley@smhs.mercy.com

BMH Sleep Disorders Center
Baptist Memorial Hospital
899 Madison Avenue
Memphis, TN 38146
Helio Lemmi, M.D.
901-227-5337 Fax: 901-227-5652

Sleep Disorders Center
Methodist Hospitals of Memphis
1265 Union Avenue
Memphis, TN 38104
Kristin W. Lester, Manager
Jim Donaldson, Supervisor
Robert Neal Aguillard, M.D.,
  Medical Director
Srinath Bellur, M.D., Assistant
  Medical Director
901-726-REST Fax: 901-726-7395

Sleep Disorders Center
Middle Tennessee Medical Center
400 North Highland Avenue
Murfreesboro, TN 37130
Timothy J. Hoelscher, Ph.D.
William H. Noah, M.D.
615-849-4811 Fax: 615-849-4833

Baptist Sleep Center
Baptist Hospital
2000 Church Street
Nashville, TN 37236
J. Michael Bolds, M.D., Director
Stephen J. Heyman, M.D.,
  Codirector
615-329-6306 Fax: 615-284-4781

Sleep Disorders Center
Centennial Medical Center
2300 Patterson Street
Nashville, TN 37203
David A. Jarvis, M.D.
Marcie T. Poe, RPSGT
615-342-1670 Fax: 615-342-1655

Sleep Disorders Center
Saint Thomas Hospital
PO Box 380
Nashville, TN 37202
J. Brevard Haynes, Jr., M.D.
Susan L. Snyder, Ph.D.
615-222-2068 Fax: 615-222-6456

TEXAS

NWTH Sleep Disorders Center
Northwest Texas Hospital
PO Box 1110
Amarillo, TX 79175
Michael Westmoreland, M.D.
John Moss, CRTT
806-354-1954 Fax: 806-351-4293

Sleep Medicine Institute
Presbyterian Hospital of Dallas
8200 Walnut Hill Lane
Dallas, TX 75231
Philip M. Becker, M.D.
Andrew O. Jamieson, M.D.
Wolfgang Schmidt-Nowara, M.D.
214-750-7776 Fax: 214-750-4621
Email: smat@sleepmed.com
Website: www.sleepmed.com

Sleep Disorders Center for Children
Children's Medical Center of Dallas
1935 Motor Street
Dallas, TX 75235
Roya Tompkins
John Herman, Ph.D.
Joel Steinberg, M.D.
214-640-2793 Fax: 214-640-7671

Sleep Disorders Center
Providence Memorial Hospital
2001 North Oregon
El Paso, TX 79902
Gonzalo Diaz, M.D., FCCP
Joseph Arteaga, RPSGT
915-577-6152 Fax: 915-577-6119

Sleep Disorders Center
Columbia Medical Center East
10301 Gateway West
El Paso, TX 79925
Gonzalo Diaz, M.D.
Jean R. Joseph-Vanderpool, M.D.
Elizabeth Baird, RPSGT
915-594-5882 Fax: 915-595-9641

Sleep Disorders Center
Columbia Medical Center West
1801 North Oregon
El Paso, TX 79902
Gonzalo Diaz, M.D.
Jean R. Joseph-Vanderpool, M.D.
Kelly Valles
915-521-1257 Fax: 915-521-1480

Sleep Consultants, Inc.
1521 Cooper Street
Fort Worth, TX 76104
Edgar Lucas, Ph.D.
C. Marshall Bradshaw, M.D.
John R. Burk, M.D.
817-332-7433 Fax: 817-336-2159
Email: sleepcon@flash.net
Website: www.flash.net/~sleepcon

Sleep Disorders Center
Department of Psychiatry
Baylor College of Medicine and VA
    Medical Center
One Baylor Plaza
Houston, TX 77030
Constance Moore, M.D., Director
Max Hirshkowitz, Ph.D.,
    Codirector
713-798-4886 and 713-794-7563
Fax: 713-798-4099 or
    713-794-7558
Email: maxh@bcm.tmc.edu or
    cmoore@bcm.tmc.edu

The Sleep Center
Spring Branch Medical Center
8850 Long Point Road
Houston, TX 77055
Todd J. Swick, M.D.
Kristyna M. Hartse, Ph.D.
713-984-3519 Fax: 713-722-3248
Email: hartsekm@intergate.com or
  tswick@ix.metcom.com

Sleep Disorders Center
Scott and White Clinic
2401 South 31st Street
Temple, TX 76508
Francisco Perez-Guerra, M.D.
254-724-2554 Fax: 254-724-2497

UTAH

Intermountain Sleep Disorders Center of
Murray Cottonwood Hospital 5770
  South
300 East
Murray, UT 84106
James M. Walker,Ph.D.
Robert J. Farney, M.D.
801-269-2015 Fax: 801-269-2948

Intermountain Sleep Disorders Center
LDS Hospital
325 8th Avenue
Salt Lake City, UT 84143
James M. Walker, Ph.D.
Robert J. Farney, M.D.
801-321-3617 Fax: 801-321-5110
Email: ldjwalke@ihc.com

Sleep Disorders Center
University Health Sciences Center
50 North Medical Drive
Salt Lake City, UT 84132
Laura Czajkowski, Ph.D.
Christopher R. Jones, M.D., Ph.D.,
  Medical Director
801-581-2016 Fax: 801-585-3249

VERMONT

No accredited center members.

VIRGINIA

Fairfax Sleep Disorders Center
3289 Woodburn Road Suite 360
Annandale, VA 22003
Konrad W. Bakker, M.D.
Marc Raphaelson, M.D.
703-876-9870

Virginia-Carolina Sleep Disorders Center
159 Executive Drive, Suite D
Danville, VA 24541
Della C. Williams, M.D., Medical
  Director
Jacalyn A. Nelson, M.D.
Mugabala B. Aswath, M.D.
William Underwood, RPSGT
Nancy Craig Williams, BS, RPSGT
804-792-2209 Fax: 804-799-8037
Email: vanc.sleep@juno.com

Sleep Disorders Center for Adults and Children
Eastern Virginia Medical School
Sentara Norfolk General Hospital
600 Gresham Drive
Norfolk, VA 23507
J. Catesby Ware, Ph.D.
Jeffery A. Scott, M.D.
Tom Bond, Psy.D.
Nancy Fishback, M.D.
757-668-3322 Fax: 757-668-2628
Email: cware@intmed1.evms.edu

Sleep Disorders Center
Medical College of Virginia
PO Box 980710 - MCV
Richmond, VA 23298-0710
Rakesh K. Sood, M.D.
804-828-1490 Fax: 804-828-1481

Sleep Disorders Center
Carilion Roanoke Community Hospital
PO Box 12946
Roanoke, VA 24029
William S. Elias, M.D.
540-985-8526 Fax: 540-985-4963

Sleep Disorders Center
Obici Hospital
1900 North Main Street
PO Box 1100
Suffolk, VA 23439-1100
Frances Davidson, RT,
   Administrative Director
Leah S. Pixley, CRTT, REEGT,
   RPSGT, Clinical Manager
Hemang Shah, M.D., Medical
   Director
757-934-4450 Fax: 757-934-4278
Email: lpixley@obici.com

Sleep Disorders Center
Virginia Beach General Hospital
1060 First Colonial Road
Virginia Beach, VA 23454
Bruce Johnson, M.D.
Yvonne Wright-Dunn, BA, RPSGT
757-481-8168 Fax: 757-496-6337
Email: ywrightdunn@
   tiaehealth.com

WASHINGTON

ARMC Sleep Apnea Laboratory*
Auburn Regional Medical Center
   Plaza One
202 North Division
Auburn, WA 98001
Julie Holdaas, RPSGT
253-804-2809 Fax: 253-735-7599

St. Clare Sleep Related Breathing Disorders Clinic*
St. Clare Hospital
11315 Bridgeport Way Southwest
Lakewood, WA 98499
Arthur Knodel, M.D.
Erin Salsbury, RPSGT
253-581-6951 Fax: 253-512-2793

Sleep Disorders Center
Southwest Washington Providence
   St. Peter Hospital
413 North Lilly Road
Olympia, WA 98506
Kim A.Chase, RPSGT
John L. Brottem, M.D.
360-493-7436 Fax: 360-493-4173

Sleep Center
Valley Medical Center
400 South 43rd Street
Renton, WA 98055
Carla J. Hellekson, M.D., FAPA
William J. DePaso, M.D., FCCP
425-656-5340 Fax: 425-656-5436
Email: erin_sheldon@valleymed.org
Website: www.valleymed.org

Richland Sleep Disorders Center
800 Swift Boulevard Suite 260
Richland, WA 99352
A. Pat Hamner, Jr., M.D.
509-946-4632 Fax: 509-942-0118
Website: www.richsleep.com

Sleep Lab*
780 Swift Boulevard, Suite 130
Richland, WA 99352
W.S. Klipper, M.D., FACCP
Claudia Havner, CRTT, NRT, PSGT
509-943-6166 Fax: 509-943-8621

Highline Sleep Disorder Center
Highline Community Hospital
14212 Ambaum Boulevard
  Southwest, Suite 201
Seattle, WA 98166
Margaret Moen, M.D., Medical
  Director
John Lovelace, RRT
Lamont Porter, RCP
206-325-7396 Fax: 206-242-2562

Virginia Mason Medical Center
Sleep Disorders Center
Virginia Mason Hospital, H10-SDC
925 Seneca Street
Seattle, WA 98101-2742
Neely E. Pardee, M.D.
Kenneth R. Casey, M.D.
Steven H. Kirtland, M.D.
Nigel J. Ball, D.Phil.
206-625-7180 Fax: 206-341-0447
Email: sdcsdc@vmmc.org

Providence Sleep Disorders Center
500 17th Avenue, Department 4W
Seattle, WA 98122
Ralph A. Pascualy, M.D.
Sarah E. Stolz, M.D.
206-320-2575 Fax: 206-320-3339
Email: lpasqualy@aol.com
Website: www.sleep.org

Seattle Sleep Disorders Center
Swedish Medical Center/Ballard
PO Box 70707
Seattle, WA 98107-1507
Gary A. DeAndrea, M.D.
Noel T. Johnson, D.O.
Richard P. Swanson, RPSGT, CRTT
206-781-6359 Fax: 206-781-6196

Sleep Disorders Center
Sacred Heart Doctors Building
105 West Eighth Avenue, Suite 418
Spokane, WA 99204
Elizabeth Hurd, RPSGT
Jeffrey C. Elmer, M.D.
509-455-4895 Fax: 509-626-4578

WEST VIRGINIA

Sleep Disorders Center
Charleston Area Medical Center
501 Morris Street
PO Box 1393
Charleston, WV 25325
George Zaldivar, M.D., FCCP
Karen Stewart, RRT, Manager
304-348-7507 Fax: 304-348-3373

PM Sleep Medicine
3803 Emerson Avenue
PO Box 4179
Parkersburg, WV 26104
Michael A. Morehead, M.D.
M. Barry Louden, M.D.
304-485-5041 Fax: 304-485-5678

WISCONSIN

Sleep Disorders Center
Appleton Medical Center
1818 North Meade Street
Appleton, WI 54911
Kevin C. Garrett, M.D.
920-738-6460 Fax: 920-831-5000

Marshfield Clinic Sleep Disorders Center
Marshfield Clinic
2655 County Highway 1
Chippewa Falls, WI 54729
Margaret Feiler
Kevin Ruggles, M.D.
715-726-4136 Fax: 715-726-4173

Luther/Midelfort Sleep Disorders Center
Luther Hospital/Midelfort Clinic
1221 Whipple Street
PO Box 4105
Eau Claire, WI 54702-4105
Donn Dexter, Jr., M.D.
David Nye, M.D.
715-838-3165 Fax: 715-838-3845

Sleep Disorders Laboratory*
Bellin Hospital
744 South Webster Avenue
Green Bay, WI 54305
John Stevenson, M.D.
Lee Kvaley, RRT
920-433-7441 Fax: 920-433-7453
Email: smercier@stvgb.org

St. Vincent Hospital Sleep Disorders Center
St. Vincent Hospital
PO Box 13508
Green Bay, WI 54307-3508
John Stevenson, M.D.
Paula Van Ert, RPSGT
920-431-3041 Fax: 920-433-8010

Wisconsin Sleep Disorders Center
Gundersen Lutheran
1836 South Avenue
La Crosse, WI 54601
Alan D. Pratt, M.D.
608-782-7300 x2870
Fax: 608-791-4466

Sleep Disorders Center
St. Mary's Hospital Medical Center
707 South Mills Street
Madison, WI 53716
Steve Dalebroux
Kathryn L. Middleton, M.D.
608-258-5266 Fax: 608-258-6176

Comprehensive Sleep Disorders Center
B6/579 Clinical Science Center
University of Wisconsin Hospitals
   and Clinics
600 Highland Avenue
Madison, WI 53792
Steven M. Weber, Ph.D.
John C. Jones, M.D.
608-263-2387 Fax: 608-263-0412
Email: smweber@macc.wisc.edu
   and jones@neurology.wisc.edu

Marshfield Sleep Disorders Center
Marshfield Clinic
1000 North Oak Avenue
Marshfield, WI 54449
Jody Scherr, RPSGT/R.EEG.T
Kevin Ruggles, M.D.
715-387-5397

**Milwaukee Regional Sleep Disorders
   Center**
Columbia Hospital
2025 East Newport Avenue,
   Suite 426Y
Milwaukee, WI 53211
Marvin Wooten, M.D.
Joni Tombari, Program Director
414-961-4650 Fax: 414-961-8712

St. Luke's Sleep Disorders Center
St. Luke's Medical Center
2801 West Kinnickinnic River
   Parkway, Suite 445
Milwaukee, WI 53215
David Arnold, RPSGT
Michael N. Katzoff, M.D.
414-649-5288 Fax: 414-649-5875
Email: dave_arnold@aurora.org

WYOMING

No accredited center members.

NOTES

CHAPTER 1 LEARNING ABOUT THE POWER OF SLEEP

1. Remark by Ronald Reagan, annual dinner, White House Correspondents' Association, April 13, 1984.

2. National Commission on Sleep Disorders Research, *Report of the National Commission on Sleep Disorders Research*, submitted to the United States Congress and to the secretary of the U.S. Department of Health and Human Services, January 1993.

3. Gallup Organization, *Sleep in America: A National Survey of U.S. Adults*, poll conducted for the National Sleep Foundation (Princeton, N.J.: National Sleep Foundation, 1995).

4. Ibid.

5. J. Allan Hobson, *Sleep* (New York: Scientific American Library, 1989).

6. Susan Page, "Travel Office Questions Ignite Clinton's Temper," *USA Today*, August 2, 1996.

7. National Commission on Sleep Disorders Research, op. cit.
8. Gallup Organization, op. cit.
9. National Sleep Foundation, *Drowsy Driving Fact Sheet* (Washington, D.C.: National Sleep Foundation, October 24, 1995).
10. National Commission on Sleep Disorders Research, op. cit.
11. Andy Pasztor, "An Air-Safety Battle Brews Over the Issue of Pilots' Rest Time," *The Wall Street Journal*, July 1, 1996.
12. James B. Maas, producer, "Sleep Alert," PBS television special, broadcast spring 1990 (Ithaca, N.Y.: Cornell University Film Unit, 1990).
13. Ibid.
14. Martin Moore-Ede, *The Twenty-Four-Hour Society* (New York: Addison Wesley, 1993).
15. National Commission on Sleep Disorders Research, op. cit.
16. William C. Dement, foreword in Elizabeth A. Mitler and Merrill M. Mitler, *101 Questions About Sleep and Dreams* (Del Mar, Calif.: Wakeful-ness-Sleep Education and Research Foundation, 1993).
17. Moore-Ede, op. cit.
18. James B. Maas, producer, "Asleep in the Fast Lane: Our 24-Hour Society," video documentary (Ithaca, N.Y.: Cornell University Psychology Film Unit, 1998).
19. Michael Segell, "The Secrets of Sleep," *Esquire*, October 1994.
20. Allan Pack, *Medical Director* (Washington, D.C.: National Sleep Foundation, 1996).
21. National Commission on Sleep Disorders Research, op. cit.
22. Ibid.
23. Ibid.
24. Ibid.
25. William C. Dement, personal communication, 1996.

CHAPTER 3 THE ARCHITECTURE AND FUNCTIONS OF SLEEP

1. J. Allan Hobson, *Sleep* (New York: Scientific American Library, 1989).
2. Thomas Ludwig, *PsychSim* (New York: Worth Publishers, Inc., 1996).
3. Hobson, op. cit.
4. Ibid.
5. A. M. Vein et al. "Physical Exercise and Nocturnal Sleep in Healthy Humans," *Human Physiology* 17 (1991): 391–397.
6. Colin Shapiro, Duncan Mitchell, Peter Bartel, and Pieter Jooste, "Slow

Wave Sleep: A Recovery Period After Exercise," *Science* 214 (1981): 1253–1254.

7. J. M. Krueger and J. A. Majde, "Sleep as a Host Defense: Its Regulation by Microbial Products and Cytokines," *Clinical Immunology and Immunopathology* (1990): 188–199.

8. Michael Irwin et al., *Journal of the Federation of American Societies for Experimental Biology* 10 (1996): 643–653.

9. *Current Research on Sleep and Dreams*, The U.S. Dept. of Health, Education, and Welfare, Public Health Service Publication No. 1389 (Washington, D.C.: U.S. Government Printing Office, 1966).

10. Ibid.

11. P. A. Maquet, D. Dive, E. Salmon, et al., "Cerebral Glucose Utilization During Sleep-Wake Cycle in Man Determined by Positron Emission Tomography and [^{18}F]2-Fluoro-2-Deoxy-D-Glucose Method," *Brain Research* 513 (1990): 136–143.

12. Carlyle Smith and Lorelei Lapp, "Increases in Number of REMs and REM Density in Humans Following an Intensive Learning Period," *Sleep* 14 (1991): 325–330.

13. O. Mandai, Alain Guerrien, P. Sockeel, Kathy Dujardin, and Pierre Leconte, "REM Sleep Modification Following a Morse Code Learning Session in Humans," *Physiology and Behavior* 46 (1989): 639–642.

14. Kathy Dujardin, Alain Guerrien, and Pierre Leconte, "Sleep, Brain Activation, and Cognition," *Physiology and Behavior* 47 (1990): 1271–1278.

15. C. Peterson, *Psychology: A Biopsychosocial Approach* (New York: Longman, 1997).

16. Hobson, op. cit.

17. Sally K. Severino, Wilma Bucci, and Monica L. Creelamn, "Cyclical Changes in Emotional Information Processing in Sleep and Dreams," *Journal of the American Academy of Psychoanalysis* 17 (1989): 555–577.

18. Peretz Lavie, *The Enchanted World of Sleep* (New Haven: Yale University Press, 1996), p. 90.

19. Ibid.

20. Harold Kudler, "The Tension Between Psychoanalysis and Neuroscience: A Perspective on Dream Theory in Psychiatry," *Psychoanalysis and Contemporary Thought* 12 (1989): 599–617.

21. Hobson, op. cit.

22. Ibid.

CHAPTER 4 SLEEP NEED AND PEAK PERFORMANCE

1. Dale M. Edgar and William C. Dement, "Evidence for Opponent Processes in Sleep/Wake Regulation," *Sleep Research* 20A:2, 1992.
2. William C. Dement and James B. Maas, *The Sleep Book for College Students* (forthcoming, 1998).
3. William C. Dement, *Some Must Watch While Some Must Sleep* (San Francisco: San Francisco Book Company, 1976).
4. Ibid.
5. Edgar and Dement, op. cit.
6. Louis Harris Poll, Louis Harris and Associates, Inc., New York, 1993.
7. William C. Dement, "Sleepiness: The Unrecognized Risk," in *Proceedings, Highway Safety Forum on Fatigue, Sleep Disorders, and Traffic Safety* (Albany, N.Y.: Institute for Traffic Safety Management and Research, 1993).
8. "Nation's Sleep Crisis Getting Worse, Say Nation's Top Docs," Gannett News Service, *Ithaca Journal*, May 25, 1995.
9. Stanley Coren, *Sleep Thieves* (New York: Free Press, 1996).
10. Ibid., p. 276.
11. Ibid., pp. 70–71.
12. National Commission on Sleep Disorders Research, *Report of the National Commission on Sleep Disorders Research,* submitted to the United States Congress and to the secretary of the U.S. Department of Health and Human Services, January 1993.
13. Robert Frost, "Stopping by Woods on a Snowy Evening," *Collected Poems, Plays and Prose* (New York: Library of America, 1995), p. 207.
14. William C. Dement, quoted in James B. Maas, producer, "Sleep Alert," PBS television special, broadcast spring 1990 (Ithaca, N.Y.: Cornell University Film Unit, 1990).
15. W. Moorcroft, *Sleep, Dreaming, and Sleep Disorders: An Introduction,* 2nd ed. (Latham, Md.: University Press of America, 1993).
16. J. Jochnowitz, "Father Released Without Bail," *Times Union* (Albany, N.Y.), June 6, 1994.
17. National Commission on Sleep Disorders Research, op. cit.
18. Paul Elias, "West Ventura County Focus: Santa Paula; Wife's Cough Led to Slaying," *Los Angeles Times,* November 3, 1995.
19. Richard Coleman, in Maas, op. cit.
20. Stanford Sleep Center, "Why Should We Care About Sleep?" pamphlet (Stanford, Calif.: Stanford Sleep Disorders Center, 1994). For more information, contact Marvin Miles, Stanford Sleep Center (415) 725-6416.

21. Henry Chu, "Americans Are Falling Asleep at the Wheel," *Rockland Journal-News* (White Plains, N.Y.), October 7, 1994.
22. Hon. John Lauber, member, National Transport Safety Board, keynote address, Association of Professional Sleep Societies, Second Annual Meeting, San Diego, Calif., June 11, 1988.
23. Andy Pasztor, "An Air-Safety Battle Brews over the Issue of Pilots' Rest Time," *The Wall Street Journal,* July 1, 1996.
24. Maas, op. cit.
25. National Commission on Sleep Disorders Research, op. cit.
26. James B. Maas, producer, "Asleep in the Fast Lane: Our 24-Hour Society," video documentary (Ithaca, N.Y.: Cornell University Psychology Film Unit, 1998).
27. Mary A. Carskadon, *Encyclopedia of Sleep and Dreaming* (New York: Macmillan, 1993).

CHAPTER 5 THE GOLDEN RULES OF SLEEP

1. Elizabeth A. Mitler and Merrill M. Mitler, *101 Questions About Sleep and Dreams* (Del Mar, Calif.: Wakefulness-Sleep Education and Research Foundation, 1993).
2. "Study: Keep Body Clock Set," *Ithaca Journal,* February 12, 1997.
3. Ibid.
4. Norman Ford, *The Sleep R$_x$.* (Englewood Cliffs, N.J.: Prentice Hall, 1994).
5. Martin Moore-Ede, *The Twenty-Four-Hour Society* (New York: Addison Wesley, 1993).
6. Lynn Lamberg, *Bodyrhythms: Chronobiology and Peak Performance* (New York: William Morrow, 1995).
7. Julia Szabo, "Sleeping Arrangements," *New Woman,* March 1995.
8. One source is the SunBox Company, 19217 Orbit Drive, Gaithersburg, MD 20879.
9. Mary A. Carskadon, *Encyclopedia of Sleep and Dreaming* (New York: Macmillan, 1993).
10. Ibid.
11. Richard Graber with Paul Gouin, *How to Get a Good Night's Sleep* (Minneapolis: Chronimed, 1995).
12. Lydia Dotto, *Losing Sleep* (New York: William Morrow, 1990).

CHAPTER 6 TWENTY GREAT SLEEP STRATEGIES

1. Richard Carlson, *Don't Sweat the Small Stuff . . . and It's All Small Stuff* (New York: Hyperion, 1997).

2. Herbert Benson, M.D., *The Relaxation Response* (New York: Avon, 1975).
3. Ibid.
4. Peter Hauri and Shirley Linde, *No More Sleepless Nights* (New York: John Wiley, 1990).
5. United Press International wire service May 21, 1990.
6. Ibid.
7. Richard Graber with Paul Gouin, *How to Get a Good Night's Sleep* (Minneapolis: Chronimed, 1995).
8. Better Sleep Council, *The Good Night Guide* (Alexandria, Va.: Better Sleep Council, 1993).
9. Graber and Gouin, op. cit.
10. "Helpful Tips for a Better Night's Sleep," advertisement, *New York Post*, May 26, 1994.

CHAPTER 7 HOW TO CREATE A GREAT BEDROOM ENVIRONMENT

1. Tips in this section were suggested by several researchers, as well as "Bedroom: Tips for a Sound Night of Sleep," *Shuteye Newsletter* (Chicago: Searle, 1996).
2. From Alecia Beldegreen, *The Bed* (New York: Stewart, Tabori & Chang, 1991).
3. Beldegreen, op. cit.
4. Ibid.
5. Ibid.
6. Jafri Mohamed, "Beds: Making the Most of the Ultimate Comfort Zone," *Straits Times* (Singapore), 6 May 1995.
7. Beldegreen, op. cit.
8. Ibid.
9. Ibid.
10. Mohamed, op. cit.
11. Courtesy of the Better Sleep Council, Box 13, Washington, DC 20044.
12. Courtesy of Pacific Coast Feather Company, 1964 Fourth Ave., Seattle, WA 98134, (206)-624-1057 (http://www.pacificcoast.com/pillows/bedpill.html).
13. Ibid.
14. "Sleeping with the Enemy?" *Snooze News* (Better Sleep Council), September 1992.
15. Beldegreen, op. cit.
16. "Test-drive Your Bed," *Snooze News* (Better Sleep Council), September 1992.

17. Beldegreen, op. cit.
18. Ibid.
19. Ibid.
20. "Test-drive Your Bed," *Snooze News* (Better Sleep Council), September 1992.

CHAPTER 8 SLEEPING PILLS AND OVER-THE-COUNTER REMEDIES

1. Richard Graber with Paul Gouin, *How to Get a Good Night's Sleep* (Minneapolis: Chronimed, 1995).
2. Peter Hauri and Shirley Linde, *No More Sleepless Nights* (New York: John Wiley, 1991).
3. Ibid.
4. Ibid.
5. Geoffrey Cowley, "Melatonin Mania," *Newsweek*, November 6, 1995; Judy Foreman, "Melatonin Bandwagon Gets Crowded," *The Boston Globe*, April 10, 1995.
6. Nava Zisapel et al., Tel Aviv University (Israel), cited in Associated Press, "Positive Results Reported in Elderly Insomnia Study," *Chicago Tribune*, August 29, 1995.
7. Saul Kassin, *Intersections: Psychology and the News* (Boston: Houghton Mifflin, 1996).
8. Ibid.
9. Foreman, op. cit.

CHAPTER 9 THE NOD TO MIDDAY NAPS

1. Richard Graber with Paul Gouin, *How to Get a Good Night's Sleep* (Minneapolis: Chronimed, 1995).
2. David McDonough, "Afternoon Delight," *TWA Ambassador*, June 1996.
3. Theda Dritchell, "The Art (and Politics) of Napping," *Cosmopolitan*, March 1996.
4. Ibid.
5. Meredith Gould, "Power Napping," *Nation's Business*, February 1995.
6. McDonough, op. cit.
7. Ibid.
8. Ibid.

CHAPTER 10 SURVIVING AS A SHIFT WORKER

1. Earl Ubell, "Do You Have Trouble Sleeping?" *Parade*, September 16, 1984.

2. Susan Cooke, "Why You Shouldn't Try to Close That Big Deal Before Dawn—or After Midnight," *Banker and Tradesman*, January 7, 1987.

3. Charles A. Czeisler, "Biological Clocks: Creating Work Schedules Based on 'Body Time,' " brochure for the Division of Endocrinology, Harvard Medical School.

4. Richard M. Coleman, "Shiftwork Scheduling for the 1990's," *Personnel* (American Management Association), January 1989.

5. Ibid.

6. Dr. Martin Moore-Ede, Institute for Circadian Physiology, brochure, Human Alertness Research Center, Cambridge, Mass.

7. Stanford Sleep Center, "Why Should We Care About Sleep," pamphlet (Stanford, Calif.: Stanford Sleep Disorders Center, 1994). For more information, contact Marvin Miles, Stanford Sleep Center, (415) 725-6416.

8. Cynthia F. Mitchell, "Firms Waking Up to Sleep Disorders," *The Wall Street Journal*, July 7, 1988.

9. Carole L. Marcus and Gerald M. Loughlin, "Effect of Sleep Deprivation on Driving Safety in Housestaff," *Sleep* 19 (1996), 10: 763–766.

10. Coleman, op. cit.

11. Ibid.

12. Jack C. Lovett, "Presentation on 'Variations in Work-Sleep Schedules: From the View of the Industrial Worker,' " *Advances in Sleep Research* 7 (1981): 325–328.

13. Coleman, op. cit.

14. American Sleep Disorders Association, Advertisement, in *Sleep* 19 (1996) no. 10.

15. Moore-Ede, op. cit.

16. Quentin Regestein and Timothy Monk, "Is the Poor Sleep of Shift Workers a Disorder?" *American Journal of Psychiatry* 148 (1991), 11: 1491.

17. Moore-Ede, op. cit.

18. Kazuya Matsumoto and Yusuke Morita, "Effects of Nighttime Naps and Age on Sleep Patterns of Shiftworkers," *Sleep* 10 (1987): 580–589.

19. Lovett, op. cit.

20. Matsumoto and Morita, op. cit.

21. Michael Irwin et al., "Partial Night Sleep Deprivation Reduces Natural Killer and Cellular Immune Responses in Humans," *Journal of the Federation of American Sleep Societies for Experimental Biology*, 10 (1996): 643–653.

22. Joseph Rutenfranz, Peter Knauth, and Dieter Angersbach, "Shift Work Research Issues," *Advances in Sleep Research* 7 (1981): 165–196.

23. Theresa R. Welker, "New Schedules Wake Up Zombies," *Industry Week*, 19 March 1990, pp. 51–52.

24. Coleman, op. cit.

25. Barbara Anderson, "Photo Therapy," *Ithaca Journal*, January 18, 1996.

26. Charles B. Inlander and Cynthia K. Moran, *67 Ways to Get Good Sleep* (New York: Walker, 1995).

27. Torbjorn Akerstedt, "Sleepiness at Work: Effects of Irregular Work Hours," in T. Monk, ed., *Sleep, Sleepiness and Performance* (New York: John Wiley, 1991), p. 144.

28. Anderson, op. cit.

29. Ubell, op. cit.

30. Cooke, op. cit.

31. Ibid.

32. Anderson, op. cit.

33. Cooke, op. cit.

34. Richard Graber with Paul Gouin, *How to Get a Good Night's Sleep* (Minneapolis: Chronimed Publishing, 1995).

35. Anderson, op. cit.

36. Cooke, op. cit.

37. Jerry E. Bishop, "Experiment Finds Work Turns Based on Inner Time Clock Superior to Abrupt Shifts," *The Wall Street Journal*, July 23, 1982.

38. "Health Benefits of Improved Scheduling," *Harvard Medical Area Focus*, 28 January 1988.

39. Richard Coleman, "What Is the Most Outdated Schedule in Continuous Operations?" *Strategic Shift* (Coleman Consulting Group), Winter 1995–96.

CHAPTER 11 REDUCING TRAVEL FATIGUE

1. R. R. Sevatson, letter to the editor, *The New York Times*, October 16, 1988.

2. Kathleen Mayes, *Beat Jet Lag* (London: Thorsons, 1991).

3. Roger Smith, Christian Guilleminault, and Bradley Efron, "Circadian Rhythms and Enhanced Athletic Performance in the National Football League," in *Sleep* 20 (1997): 362–365.

4. Charles Ehret and Lynne Waller Scanlon, *Overcoming Jet Lag* (New York: Berkley, 1983).

5. Ibid.

6. Mayes, op. cit.

7. Dan Oren, Walter Reich, Norman Rosenthal, and Thomas Wehr, *How*

to Beat Jet Lag: A Practical Guide for Air Travelers (New York: Henry Holt, 1993).

8. R. Curtis Graeber, "Jet Lag and Sleep Disruption," in *Principles and Practice of Sleep Medicine*, ed. M. H. Kryger et al. (Philadelphia: W. B. Saunders, 1994).

9. Mayes, op. cit.

10. Ibid.

11. Richard M. Coleman, *Wide Awake at 3:00 A.M.* (New York: W. H. Freeman, 1986).

12. Ehret and Scanlon, op. cit.

13. Mayes, op. cit.

14. "Tips for Overcoming Jet Lag," brochure, Upjohn Company, Kalamazoo, Mich., 1983.

15. Thanks to researchers K. Mayes, W. C. Dement, R. Graber, P. Gouin, C. Ehret, L. Scanlon, D. Oren, W. Reich, N. Rosenthal, and T. Wehr.

16. Catherine Chetwynd, "Down to Earth Advice for Those up in the Air," *London Times*, November 3, 1995.

17. Mayes, op. cit.

18. Available from the SunBox® Company, 19217 Orbit Drive, Gaithersburg, MD 20879.

19. "Sleep and the Traveler," brochure, National Sleep Foundation and Hilton Hotels, 1995.

20. Nancy H. Butler, "National Forum on Sleeplessness and Crashes Staged in Washington, D.C.," *The National Sleep Foundation Connection*, Winter 1995.

21. Sandy Rovner, "The Danger in Driving Drowsy," *The Washington Post*, January 10, 1995.

22. Roger Fritz, *Sleep Disorders: America's Hidden Nightmare* (Nashville: National Sleep Alert, Inc. 1993).

23. Rajiv Chandrasekaran, "Eighteen Wheels and Forty Winks," *The Washington Post*, June 22, 1995.

24. Fritz, op. cit.

25. Richard Cole, "Study Finds Sleep Apnea Raises Risk of Car Crash," Associated Press, 22 May 1997.

26. Mary A. Carskadon and Joan Mancuso, "How to Avoid Falling Asleep While Driving," in *Drive Alive . . . Drive Alert!* (Rochester, Minn.: Association of Professional Sleep Societies, 1987).

27. Jane Brody, "Drowsy Driving Can Be Deadly," *Star Tribune*, December 25, 1994.

28. Butler, op. cit.

29. Matthew Wald, "Saying Driver Faked Log Entries, U.S. Faults Trucking Firm in Fatalities," *The New York Times*, November 14, 1994.
30. Henry Chu, "Americans Are Falling Asleep at the Wheel," *Rockland Journal-News* (White Plains, N.Y.), October 7, 1994.
31. Brody, op. cit.
32. Dr. Thomas Roth, personal communication, Henry Ford Hospital, Detroit, June 1993.
33. Doug Levy, "A Wake-Up Call for Road Safety," *USA Today*, June 25, 1996.
34. Dr. William C. Dement, personal communication, June 1993.
35. "National Forum on Sleeplessness and Crashes Staged in Washington, D.C."
36. Ibid.
37. Mary C. Hickey, "Asleep at the Wheel," *Ladies Home Journal*, December 1994.
38. Ibid.
39. *Wake Up!* (Washington, D.C.: AAA Foundation for Traffic Safety, 1996).
40. The sleep and safety tips in this chapter are a compilation of information made available by the National Highway Traffic Safety Administration, the Association of Professional Sleep Societies, Mary Carskadon, Joan Mancuso, the American Automobile Association Foundation for Traffic Safety, and the Better Sleep Council.

CHAPTER 12 AVOIDING FAMILY SLEEP TRAPS

1. Stanley Coren, *Sleep Thieves* (New York: Free Press, 1996).
2. Paula Siegel, "Helping Your Child and You Get a Good Night's Sleep," *Good Housekeeping*, September 1995.
3. Thanks to Denton Wilson of Nepean, Ontario, for this catchy phrase.
4. Better Sleep Council, *The Good Night Guide* (Alexandria, Va.: Better Sleep Council, 1993).
5. *The Treatment of Sleep Disorders of Older People*, NIH Consensus Development Conference, March 26–28, 1990; 8(3): 1–22.
6. Ibid.
7. Richard Graber with Paul Gouin, *How to Get a Good Night's Sleep* (Minneapolis: Chronimed Publishing, 1995).
8. Jack Cheevers, "Quieting Noisy Nursing Homes; Elderly Study Says Disruptions are Affecting Residents' Sleep at Four Valley Sites. Researchers Hope Findings Will Encourage Solutions," *Los Angeles Times*, December 3, 1995.

9. Nava Zisapel et al., Tel Aviv University (Israel), cited in Associated Press, "Positive Results Reported in Elderly–Insomnia Study," *Chicago Tribune*, August 29, 1995.

10. D. Stretch et al., University of Leicester (U.K.), cited in Associated Press, "Elderly and Can't Sleep? Try Scent of Lavender," *The New York Times*, September 13, 1995.

11. Graber and Gouin, op. cit.

12. Zisapel, op. cit.

13. "When Counting Sheep Doesn't Help Get You to Sleep," *Kiplinger's Retirement Report*, October 1994.

14. Available from the SunBox® Company, 19217 Orbit Drive, Gaithersburg, MD 20879.

15. Mary A. Carskadon, *Encyclopedia of Sleep and Dreaming* (New York: Macmillan, 1993).

16. Ibid.

17. Better Sleep Council, "Women: Doing More and Sleeping Less," *Snooze News*, May 1993 (Alexandria, Va.: Better Sleep Council).

18. Mark R. Rosekind, "The Epidemiology and Occurrence of Insomnia," *Journal of Clinical Psychiatry* 53 (1992): 4–6.

19. Eva Lindberg, Christer Janson, Thorarinn Gislason, Eythor Björnsson, Jerker Hetta, and Gunnar Bowman, "Sleep Differences in a Young Adult Population: Can Gender Differences Be Explained by Differences in Psychological Status?" *Sleep* 20 (1997): 381–387.

20. N. G. Kutner et al., "Older Adults' Perceptions of their Health and Functioning in Relation to Sleep Disturbance," *Journal of the American Geriatric Society* 42 (1992): 757–762.

21. Carlos H. Schenck and Mark Mahowald, "Review of Nocturnal Sleep-Related Eating Disorders," *International Journal of Eating Disorders* 15 (1994): 343–356.

22. D. W. Wetter and T. B. Young, "The Relation Between Cigarette Smoking and Sleep Disturbance," *Preventative Medicine* 23 (1994): 328–334.

23. R. Levin, "Sleep and Dreaming Characteristics of Frequent Nightmare Subjects in a University Population," *Dreaming: Journal of the Association for the Study of Dreams* 4 (1994): 127–137.

24. Calvin S. Hall and Robert L. Van de Castle, *The Content Analysis of Dreams* (New York: Appleton-Century-Crofts, 1966).

25. Carlos H. Schenck and Mark Mahowald, "Polysomnographic, Neurologic, Psychiatric, and Clinical Outcome Report on Seventy Consecu-

tive Cases with REM Sleep Behavior Disorder (RBD)," *Cleveland Clinic Journal of Medicine* 57 (1990): supp. s9–23.
26. Carskadon, op. cit.
27. Ibid.
28. Barbara L. Anderson, "Primary Orgasmic Dysfunction: Diagnostic Considerations and Review of Treatment," *Psychological Bulletin* 83 (1983): 105–133.
29. Stuart A. Lewi and Maureen Burns, "Manifest Dream Content: Changes with the Menstrual Cycle," *British Journal of Medical Psychology* 48 (1975): 375–377.
30. Mary Brown Parlee, "Changes in Moods and Activation Levels During the Menstrual Cycle in Experimentally Naive Subjects," *Psychology of Women Quarterly* 7 (1982): 119–131.
31. Carskadon, op. cit.
32. Ibid.

CHAPTER 13 INSOMNIA AND BEYOND

1. Gallup Organization, *Sleep in America: A National Survey of U.S. Adults,* poll conducted for the National Sleep Foundation (Princeton, N.J.: National Sleep Foundation, 1995).
2. Thomas Roth, "Social and Economic Consequences of Sleep Disorders," *Sleep* 19 (1996): 46–47.
3. Paul Elias, "West Ventura County Focus: Santa Paula; Wife's Cough Led to Slaying, Jury Told," *Los Angeles Times,* November 3, 1995.
4. National Commission on Sleep Disorders Research, *Report of the National Commission on Sleep Disorders Research,* submitted to the United States Congress and to the Secretary of the U.S. Department of Health and Human Services, January, 1993. Dept. of Health and Human Services. Publication no. 92 (Washington, D.C.: U.S. Government Printing Office, 1992).
5. William C. Dement, testimony before House Appropriations Committee hearings on funding for sleep disorders research (May 22, 1985).
6. National Commission on Sleep Disorders Research, op. cit.
7. Gallup Organization, op. cit.
8. Diagnostic Classification Steering Committee, ed. Michael J. Thorpy, *The International Classification of Sleep Disorders: Diagnostic and Coding Manual* (Rochester, Minn.: American Sleep Disorders Association, 1990).
9. National Sleep Foundation, *The Nature of Sleep* (Washington, D.C.: National Sleep Foundation, 1995).

10. J. R. Goldberg, ed. *The Pharmacological Management of Insomnia* (Washington, D.C.: National Sleep Foundation, 1996).
11. Ibid.
12. Philip R. Westerbrook, "Apnea," in *Encyclopedia of Sleep and Dreaming* (New York: Macmillan, 1993).
13. Case history courtesy of Dr. Mark Ivanick, St. Joseph's Hospital Sleep Disorders Center, Elmira, N.Y.
14. Roger Fritz, *Sleep Disorders: America's Hidden Nightmare* (Naperville, Ill.: National Sleep Alert, Inc., 1993).
15. *International Classification of Sleep Disorders,* op. cit.
16. National Commission on Sleep Disorders Research, op. cit.
17. "It's Time to Wake Up to the Importance of Sleep Disorders," *Journal of the American Medical Association* 269 (1993).
18. *International Classification of Sleep Disorders,* op. cit.
19. Fritz, op. cit.
20. Ibid.
21. Ibid.
22. Barbara Nachman, "Snooze You Can Use: Stifling a Partner's Snores," *USA Today,* April 5, 1994.
23. *International Classification of Sleep Disorders,* op. cit.
24. T. Scott Johnson, M.D., and Jerry Halbustadt, *Phantom of the Night* (Cambridge, Mass.: New Technology Publishing, Inc., 1995).
25. Nachman, op. cit.
26. The Sleep Well, Stanford University website (http://www-leland. stanford.edu/~dement/snoring.html).
27. Norman Ford, *The Sleep R_x* (Englewood Cliffs, N.J.: Prentice Hall, 1994).
28. *International Classification of Sleep Disorders,* op. cit.
29. Ibid.
30. Ibid.
31. Ibid.
32. Case history courtesy of Dr. Mark Ivanick, St. Joseph's Hospital Sleep Disorders Center, Elmira, N.Y.
33. Ibid.
34. National Sleep Foundation, op. cit.
35. *International Classification of Sleep Disorders,* op. cit.
36. Ibid.
37. Case history courtesy of Dr. Mark Ivanick, St. Joseph's Hospital Sleep Disorders Center, Elmira, N.Y.
38. *International Classification of Sleep Disorders,* op. cit.
39. Ibid.

40. Andrew Jamieson and Philip M. Becker, "Management of the 10 Most Common Sleep Disorders," *American Family Physician* 45 (1992): 1262–1268.
41. *International Classification of Sleep Disorders*, op. cit.
42. Ibid.
43. Ibid.
44. Ibid.
45. Ibid.
46. Ibid.
47. Case history courtesy of Dr. Mark Ivanick, St. Joseph's Hospital Sleep Disorders Center, Elmira, N.Y.
48. Case history courtesy of Dr. Harly Greenberg, Sleep-Wake Disorders Center, Long Island Jewish Medical Center.
49. Ford, op. cit.
50. Ibid.
51. Ibid.
52. *International Classification of Sleep Disorders*, op. cit.
53. Ibid.
54. Fritz, op. cit.
55. Ibid.
56. Monique Roffey, " 'But I Was Asleep Through the Whole Thing, Your Honour': Sleepwalking Isn't Always Comic; People Can Get Hurt," *The Independent*, September 25, 1994.
57. John F. Simonds and Humberto Parraga, "Prevalence of Sleep Disorders and Sleep Behaviors in Children and Adolescents," *Journal of the American Academy of Child Psychiatry* 21 (1982): 383–388.
58. Ibid.
59. Ibid.
60. *International Classification of Sleep Disorders*, op. cit.
61. Ibid.
62. *Sleep Disorders*, U.S. Dept. of Health and Human Services publication no. (ADM) 87-1541 (Washington, D.C.: U.S. Government Printing Office, 1987).
63. Fritz, op. cit.
64. *International Classification of Sleep Disorders*, op. cit.
65. Ibid.
66. *Sleep Disorders*, op. cit.

INDEX

Horne, Marilyn, 92
hospitals, 11, 52, 133, 135,
    170–71
hotels, 151–52
hot flashes, 175
humidity, 104
humor, 60, 85, 97, 189
hyperactivity, 117
hypersomnia, 73, 170, 193–94
    "healthy," 194
    idiopathic, 194
    posttraumatic, 194
    recurrent, 193
hypertension, 30*n*, 91, 187–88
hypnagogic hallucinations, 31,
    188, 189, 190
hypnotic drugs, 170, 184

imipramine, 201
immune systems, xvi, 26–27, 34,
    60, 78, 120, 138
impotence, 36, 174
income loss, 13
infants, 28, 75, 77, 84, 102, 128,
    146, 148
    caring for, 163–66
    crying of, 165
    death of, 13, 201–2
    minimizing nighttime activity
        and feedings of, 164
    sleeping pill addictions of, 117
    sleep schedules for, 165
    teething of, 165
infertility, 10
information, 27, 37–40, 54
    see also memory; thinking
insomnia, 183–84
    causes of, 76, 78, 84, 87–88,
        117, 119, 145, 170–71, 183,
        202–3
    chronic, 116, 124, 184
    economic costs of, 180
    in elderly people, 120, 127,
        170–73
    short-term, 83, 116, 119, 183
    Sunday-night, 75–76, 77

symptoms of, 84
treatment of, 12, 115–19, 141,
    168, 170, 171–73, 183–84
Institute for Circadian Physiology,
    136, 141
Institute for Traffic Safety, 156
*International Classification of Sleep
    Disorders Diagnostic and Coding
    Manual, The,* 182
irritability, 60, 117, 128, 137, 146,
    166, 188
Irwin, Michael, 34, 137

Jansen, Dan, 72
Japan, 55
jet lag, 10, 47, 76, 115, 120, 131,
    134, 143–53, 195
    combating of, 147–53
    detrimental effects of, 144–45,
        145–46
    factors increasing
        susceptibility to, 146–47
    overcoming of, 143–45
    personality characteristics and,
        147
    symptoms of, 145–46
Johnson, Lyndon, 4
Joplin, Janis, 3

Kennedy, John F., 124
Kenny, Elizabeth, 34
Kerwin, Joe, 113
Kleine-Levin Syndrome, 193
Kleitman, Nathaniel, 28–29
Klonopin, 192, 198
Koh, Eunsook, 88

lavender oil, 171
Lavie, Peretz, 30*n*
Lawrence, D. H., 205
laziness, 5, 80, 82, 128, 129, 180
Leno, Jay, 3
Leonardo da Vinci, 128
lethargy, 60, 129, 168
life span, 15, 137
lifestyles, 123

light, 134
    artificial, 7, 26
    blocking of, 103, 140
    daylight full-spectrum,
        138–40, 150, 195–96
    electric, 7
    experiments with, 47
    exposure to, 47, 81, 138–40,
        150, 173, 195–96
    natural, 7, 47, 151, 152, 173
    treatment with, 81, 138–40,
        150, 195–96
Linde, Shirley, 89
Louganis, Greg, 144
L-tryptophan, 184
lullabies, 167

Mahowald, Mark, 81
Marine Corps, U.S., 57
Marx, Groucho, 106
masturbation, 92
mattresses, 108–12
Mayo Clinic Insomnia Program,
    89
McCarthy, Ann, 156
medical schools, 181, 182
medical workers, 11, 52, 133,
    135, 170–71
meditative relaxation, 84–85, 94,
    127
melatonin, 47, 119–20, 150, 171,
    172, 184
memory, 6–7, 72, 168
    loss of, 145
    organization and
        reorganization of, 37, 40–41
    prioritization of, 41
    storage and retention of,
        37–40, 41, 78
memory traces, 37
menopause, 175
menstrual disorders, 146
menstruation, 175
metabolism, 33
    cerebral glucose, 37
microsleeps, 60, 155

Midgow, Jeffrey, 126
mind games, 95
minerals, 89
Miyazawa, Kiichi, 10
Monday Night Football, 144
monosodium glutamate (MSG),
    88
mood, xvi, 6–7, 8, 15, 51, 55, 60,
    64, 65, 72, 85, 97, 126, 137,
    150, 168, 184, 202
Moore-Ede, Martin, 10, 11, 134,
    141
mothers
    breastfeeding, 164
    infant care by, 77, 128,
        163–66
    working, 7
motor coordination, 60, 72,
    145
Multiple Sleep Latency Test
    (MSLT), 62–64
muscles, 35, 66
    relaxation of, 31, 32–33, 85,
        86, 94–95, 153, 188–89
muscle tonus, 29, 31, 62, 124,
    188–89
music, 91, 95, 103, 130, 161

naps, 64, 76, 78, 123–32, 137,
    138, 166, 169–70, 189, 190
    famous believers in, 4, 121, 124
    guidelines for, 130–32, 173
    irregular, 126, 128
    off the road, 159
    power, 129–30
    prophylactic, 127–28
    statistics on, 123–24
    tendency for, 124–25
    timing and duration of, 79,
        123, 126–27, 130–31, 173
    workplace, 128–32
narcolepsy, 62, 188–91
    symptoms of, 188–90
    treatment of, 190
Narcolepsy Network, 191
nasal polyps, 187

National Aeronautics and Space
    Administration (NASA),
    113–14, 135–36
National Commission on Sleep
    Disorders Research, 13, 55,
    181
National Football League, 144
National Highway Safety
    Administration, 154
National Sleep Foundation, 9, 12,
    160, 184, 190
National Transportation Safety
    Board, 9, 58
neckaches, 106
nervous system, 35, 36, 102, 126
neural networks, 37, 39, 40
neurons, 37, 39
neurotransmitters, 41–42
nicotine, 89
nightclothes, 105, 167
nightlights, 151, 165
nightmares, 103, 167, 170, 174
night owls, 80–82, 147
night sweats, 103
noise, 102–3, 139, 151, 171
No More Sleepless Nights (Hauri and
    Linde), 89
non-rapid-eye-movement
    (NREM) sleep, 29, 37, 89,
    90, 174
norepinephrine, 41–42
nuclear accidents, 11
nursing homes, 170–71
nutrition, 14, 87–89
Nytol, 118

Oklahoma, University of, Health
    Sciences Center at, 88
orgasmic dysfunction, primary,
    174–75
orthopedists, 106
osteoporosis, 170

painkillers, 171
palpitations, 52
panic disorders, 184, 203

paranoia, 8
parasomnias, 196–201
parasympathetic nervous system,
    36
Parkinson's disease, 184
Parlodel, 192
peak performance
    preparing the mind for, 14,
        17, 25, 27, 42–43, 54,
        69–114, 163
    REM sleep and, 37–43
    sleep need and, xv, xvi, xvii,
        5, 7–8, 14, 33–34, 42–43,
        45–67
Peak Performance Sleep Logs,
    96–98, 209–15
penile erections, 35, 36, 174
Pennsylvania, University of, 125,
    154
perception, 6–7
performance, 7–11, 13
    improvement of, 72–73,
        96–98, 132
    sleep deprivation and, xvi, 38,
        51, 55–61, 64, 65
periodic limb movement disorder,
    193
persecution, delusions of, 55
Philadelphia Police Department,
    141
physical examinations, 13–14
physicians, 11, 13–15, 91, 97,
    116, 117–18, 120, 135, 140,
    156, 165, 172, 181, 184,
    198, 201
pillows, 106–8, 130, 149, 151
pineal gland, 120
pituitary gland, 33
plane crashes, 9–10, 56
Plante, Bill, 8
police, 133, 135, 141
polygraphs, xiv, 29–30, 62
Poor Richard's Almanac (Franklin),
    69
positron emission tomography
    (PET), 37, 38

pregnant women, 75, 91, 116–17, 175, 192
productivity, xvi, 6–7, 11, 13, 14, 15, 55, 61, 129, 130, 179–80
    shift work and, 134–36, 138, 141–42
progressive muscle relaxation (PMR), 94
prostate problems, 77, 170
protein, 87, 88
psychiatric disorders, 8, 137, 184, 202–3
psychoses, 184, 202

quality of life, xiii, xvi, xvii, 7–8, 12, 15

radios, 103, 161
rapid eye movement (REM), xv, 28–29
    see also REM sleep
reaction time, 6–7, 60–61, 72, 145
reading, 92, 95, 167
Reagan, Ronald, 4, 124
recreation, 7–8, 85–86
Reeves, James, 56
relaxation response, 84–85
relaxation techniques, 84–85, 86, 91–92, 94–95, 127, 192, 200
REM rebound effect, 37
REM sleep, 34–43, 124, 188, 189, 190
    ascending to, 34–36
    cycles of, 42–43
    deprivation of, 42–43, 89, 90, 145, 149
    dreaming in, xv, 28–29, 34–36, 39, 40–41, 42, 89
    increases in, 38, 42
    learning and, 37, 38, 40–41, 42–43, 168
    memory and, 37–41, 42–43
    peak performance and, 37–43
REM sleep behavior disorder, 196–98

respiration, xiv, 26, 29, 30, 31, 35, 62
    depression of, 97, 117
    temporary cessation of, 12–13, 90–91, 154, 184–88, 187
rest, 26, 56
restless legs syndrome, 191–92
Restless Legs Syndrome Foundation, 192
Rich, James, 57
Roehrs, Timothy, 54
Roth, Thomas, 54, 64, 73, 156, 179–80
Roy, Gabrielle, 40

safety, 6–7, 197
San Diego Veterans' Hospital, 137
schizophrenia, 202
Schnelle, John, 171
security, 104, 197
sedatives, 184
sedentary activities, 85–86, 127
Selye, Hans, 84
separation anxiety, 166
serotonin, 41–42, 88
sexual activity, 36, 80–81, 91, 92, 136
"Shadows" (Lawrence), 205
sheep, counting of, 95
sheets, 105–6
shift work, 133–42
    effective schedules for, 141–42
    health and, 136–37, 141
    mistakes and accidents in, 135–36, 141
    productivity and, 134–36, 138, 141–42
    seven swing, 134
    sleep deprivation and, 10–11, 47, 75, 115, 134–37
    sleep strategies for, 138–40
sick days, 137
Sidney, Philip, 23
siestas, 125, 129
Silki, 109

Sinemet, 192
sleep
    architecture and functions of,
        25–43
    arousal from, 31–32, 59, 62,
        171
    continuity of, 77–78, 123,
        164
    deep, xiv, 13, 27–28, 29,
        33–34, 76, 77, 85, 86, 91,
        126, 145, 174
    definitions of, 26
    disturbances of, 77–78, 84, 95,
        96, 102–3, 115, 145
    establishing and maintaining
        schedules for, 15, 48,
        52–53, 72–77, 96–98, 104,
        168
    feigning of, 26
    gender differences and, 5, 35,
        36, 173–75, 186, 197
    golden rules of, 72–80, 97, 164
    inducement of, 12, 45–50, 88,
        115–19, 141, 168–69,
        171–73
    knowledge and education
        about, xv, 4–5, 14–15,
        17–18, 22, 25–27, 132,
        181–82
    learning in, 40, 41–42
    opponent-process model of
        wakefulness and, 45–50
    power of, 6–7, 14–15, 17–18,
        54–55, 72–73, 96, 163
    rhythms and stages of, xiv,
        xvii, 10, 26, 28–43, 76–77,
        125, 129, 134
    shallow, 27–28, 29, 96
    suggested readings and
        videotapes on, 217–20
    "twilight," 34–35
    value placed on, xvi, xvii, 7–8,
        14, 54–55, 96
    varied physiological activity
        in, xiv–xv, 5–6, 15, 26,
        27–43, 47

sleep aids, 119–20, 150
"Sleep Alert," 9–10
sleep apnea, 12–13, 62, 117
    accident rates and, 154, 156,
        185–86
    alcohol and, 90–91
    central, 186
    definitions of, 12, 187
    detrimental effects of, 12–13,
        154, 156, 185–86
    health dangers of, 12, 90–91,
        186–88
    obstructive, 184–88, 194
    repetitive breathing pauses in,
        12, 90–91, 154, 184–88,
        187
    testing for, 187
    treatment of, 186–87, 187
sleep deprivation, 50–53
    alcohol and, 90, 91
    causes of, 7–8, 54–55
    conditioning and, 6, 7–8, 75
    coping with, 4–5, 65–67,
        72–100, 121–61
    cumulative effects of, 52–55,
        57, 64–65, 78–79, 136
    debt incurred in, 46, 48, 50,
        52–53, 54–55, 57, 62–65,
        78–80, 125, 126, 129, 147,
        156
    detrimental effects of, xvi,
        7–12, 22, 34, 41–43, 53–67,
        129, 144–45, 155, 156
    extent of, 5
    financial costs of, 9, 10–11,
        13–14, 58, 129, 135–36
    growing trend in, xvi, 7–8
    performance and, xvi, 38, 51,
        55–62, 64
    repayment of, 34, 48, 53,
        54–55, 77, 78–80, 127
    signs of, 19, 50–51, 59–65
    statistics on, xvi, 8–11, 14,
        52
    testing for, 5–6, 17–22, 62–65
    violence and, 57

Wisconsin, University of, 154
work
    commutation to, 7
    early-morning, 77
    emergency, 128
    falling asleep at, 9–10, 11, 13,
        56–59, 126–27, 128–32,
        134, 141
    increased hours spent at, 7–8,
        10–11, 72–73, 136,
        156–57

night, 11, 133, 136, 141
satisfaction with, 136
sleep deprivation and, 5, 7–8,
    10–11, 48, 75, 116, 134–36
    see also shift work
World War II, 102, 121, 124,
    133

yawning, 66–67, 159
yoga, 94
Young, Terry, 154

ABOUT THE AUTHOR

DR. JAMES B. MAAS is a professor of psychology and the Stephen H. Weiss Presidential Fellow at Cornell University, where he teaches introductory psychology to 1,500 students each year in the nation's largest single lecture class and conducts research on the psychophysiology of sleep.

Dr. Maas has held a Fulbright Senior Professorship to Sweden, has been a visiting professor at Stanford University, and was president of the American Psychological Association's Division on Teaching. He has received the Clark Award for Distinguished Teaching at Cornell and the American Psychological Association's Distinguished Teaching Award.

He is also an award-winning filmmaker who has produced television specials for PBS in the United States, for the BBC in England, for the CBC in Canada, and for Dutch, Danish, and Swedish national television. Dr. Maas is one of the nation's most sought-after speakers and seminar leaders for organizations and businesses, and he makes frequent television appearances. For information on his multimedia keynote addresses and seminars on the concepts discussed in *Power Sleep*, contact Maas Presentations, 6 Sunset West, Ithaca, NY 14850 (phone and fax: 607-347-6561).